The Urban Book Series

The Urban Book Series is a resource for urban studies and geography research worldwide. It provides a unique and innovative resource for the latest developments in the field, nurturing a comprehensive and encompassing publication venue for urban studies, urban geography, planning and regional development.

The series publishes peer-reviewed volumes related to urbanization, sustainability, urban environments, sustainable urbanism, governance, globalization, urban and sustainable development, spatial and area studies, urban management, transport systems, urban infrastructure, urban dynamics, green cities and urban landscapes. It also invites research which documents urbanization processes and urban dynamics on a national, regional and local level, welcoming case studies, as well as comparative and applied research.

The series will appeal to urbanists, geographers, planners, engineers, architects, policy makers, and to all of those interested in a wide-ranging overview of contemporary urban studies and innovations in the field. It accepts monographs, edited volumes and textbooks.

Indexed by Scopus.

Fabio Naselli · Edmond Manahasa · Egin Zeka ·
Kreshnik Merxhani
Editors

Self-Sufficient Cities and Sustainable Habitats

Perspectives and Case Studies

Springer

Editors
Fabio Naselli
Department of Architecture
Epoka University
Tirana, Albania

Edmond Manahasa
Department of Architecture
Epoka University
Tirana, Albania

Egin Zeka
Department of Architecture
Epoka University
Tirana, Albania

Kreshnik Merxhani
Department of Architecture
Epoka University
Tirana, Albania

ISSN 2365-757X ISSN 2365-7588 (electronic)
The Urban Book Series
ISBN 978-3-032-10336-9 ISBN 978-3-032-10337-6 (eBook)
https://doi.org/10.1007/978-3-032-10337-6

This Springer imprint is published by the registered company Springer Nature Switzerland AG
The registered company address is: Gewerbestrasse 11, 6330 Cham, Switzerland

Contents

1 **Introducing Sustainable Habitats and Self-sufficient Cities** 1
Egin Zeka and Kreshnik Merxhani

2 **Urban Regeneration Through System Planning and a Multiscalar Approach for a Self-sufficient City: The Case of Santa Maria Capua Vetere in Italy** 11
Antonio Acierno and Ivan Pistone

3 **The Urban Project of the Testaccio District in Rome: How a Compact Fabric Can and Must Accommodate Strategic Transformation Initiatives for a Sustainable Habitat Progress** 39
Cinzia B. Bellone and Paolo Trevisani

4 **'Cartographies' of the Landscape of the Interwar Linear Park of Fokionos Negri Street in Athens Through Literature and Art** 57
Georgia Eleftheraki, Maria Markou, and Anna Maria Visilia

5 **Abandoned Factories and Industrial Landscape in Albania** 83
Mariateresa Giammetti and Kreshnik Merxhani

6 **Tirana-Next: Exploring Potential Strategies for the City's Sustainable Third Landscape** 95
Fabio Naselli and Mariateresa Giammettii

7 **The Relationship Between Heritage Buildings and Their Urban Unit—Reinterpreting the Role of the Urban Context in Historical Habitats Facing Rapid Transformations, Case of Tirana** 113
Marsela Plyku Demaj, Skender Luarasi, and Blerim Nika

8 **Laic and Religious Areas in the Small Town of Vigonza (Italy). Examples of Urban Mending Projects** 125
Enrico Pietrogrande, Alessandro Dalla Caneva, and Massimo Mucci

9 Walking in Vital Space: Timeline of Vitality in the Inner City of Urmia 143
Momen Foadmarashi and Francisco Serdoura

10 How Walking-Streets Promoting Sightseeing: Effect of Walking Routes on the Tourist Places of Hamadan Inner City ... 161
Solmaz Asadpour, Momen Foadmarashi, and Francisco Serdoura

11 Unveiling the Long-Term Value of Green Buildings for Self-Sufficient Units: A Life Cycle Cost and Budget Analysis in the Albanian Landscape 177
Julinda Keçi and Geron Rakipaj

12 Lessons Learned from Sustainable Habitats and Self-sufficient Cities in the Euro-Mediterranean Context 195
Edmond Manahasa and Fabio Naselli

Index 201

Chapter 1
Introducing Sustainable Habitats and Self-sufficient Cities

Egin Zeka and Kreshnik Merxhani

Abstract Recent developments have resulted in some significant worldwide concerns. Changes in the climate, natural disasters, pandemics, and wars have all altered the local, regional, and global environment. Migration, supply of energy, economic, and climatic issues are prompting many countries to evaluate and consider new innovative approaches. It is clear that dealing with these difficulties needs more than a political approach; it requires a multidisciplinary and cross-sectoral approach. Urban planning, architecture, engineering, and design all can play important roles in problem-solving. In addition, dealing with the built environment requires a more comprehensive, trans-disciplinary approach. A sustainable environment and self-sufficient city seek to balance urban development and ecological consumption, assuring the long-term viability of both human and natural systems. This shift is based on the combination of green infrastructure, renewable energy sources, and waste management systems that reduce environmental effects. A self-sufficient city seeks to produce its "own" energy, food, and resources on-site, minimizing reliance on outside supply and lowering the carbon footprint. Key aspects include the utilization of solar, wind, and other forms of renewable energy, urban farming, water recycling, and energy-efficient buildings. Furthermore, sustainable transportation options such as public transportation, cycling, and electric vehicles are given priority to minimize emissions. Cities must incorporate green spaces that promote biodiversity and air quality while also providing leisure opportunities for people. A self-sufficient city promotes community participation by encouraging inhabitants to participate in sustainability activities and decision-making processes. By balancing economic, social, and environmental concerns, such cities can lower their ecological footprint, minimize the consequences of climate change, and improve the quality of life for their residents, thereby establishing a paradigm for future urban growth.

E. Zeka (✉) · K. Merxhani
Department of Architecture, Epoka University, Tirana, Albania
e-mail: ezeka@epoka.edu.al

K. Merxhani
e-mail: kmerxhani@epoka.edu.al

F. Naselli et al. (eds.), *Self-Sufficient Cities and Sustainable Habitats*, The Urban Book Series, https://doi.org/10.1007/978-3-032-10337-6_1

Keywords Sustainable environment · Self-sufficient city · Balanced urban development · Ecological consumption · Long-term viability · Minimized emissions · Carbon footprint · Quality of life

1.1 Introduction

The issue of urbanization and its relation to natural resources will be a fundamental and real challenge regarding sustainability in the future. As commonly put forward by the UN, by 2050, it is estimated that around 70% of the world's population will live in urban areas (UN, nd.) In this situation, problems of managing natural resources will be even more evident in local and global contexts (Vargas et al., 2019). This is directly connected to the character of cities as main regions of resource consumption and centers of waste production (Medeiros, 2025). For this reason, to reduce the environmental impact of cities, urban resources should be managed more efficiently. The relationship between cities, people, and nature needs to be rethought and urban zones must be considered as agents of adaptability, that use resources efficiently and have a regenerative nature (Medeiros, 2024; (urbanNEXT, nd.).

The attention of many countries and institutions toward environmental issues and sustainable development has been increased and many initiatives have been taken. In the year 2000, the UN member countries had an agreement on Millennium Development Goals [MDGs] which was followed by a more comprehensive strategy in 2015 that defined the Agenda for 2030 with 17 new Sustainable Development Goals [SDGs] (UN, UNDP, nd.) These strategies and goals cover a wide range of social, economic, and environmental topics. In the same year [November 2015], 196 parties of the UN drafted a strategy for climate change that is known as the Paris Agreement (EUCO, n.d.). Along the same line, in 2019 the EU countries launched the European Green Deal by defining a set of policy initiatives for a 'green transition' and 'climate neutrality' for the target year 2050. According to European Council, "The European Green Deal aims to create a cleaner, healthier and climate-neutral Europe by transforming the way we produce and consume" (EUCO, n.d.). This is a clear statement of the European Union, supported by new legislation that aims to deal with climate change, social equality, economic prosperity reconciliation, and the use of natural resources in a sustainable way.

To achieve the objectives of the Green Agenda, the Regional and city planning approaches need to be re-dimensioned by embracing the concept of the self-sufficient city. This concept of self-sufficiency supported by new policies, programs, and projects would have a great contribution to reach a sustainable environment and resilient cities and regions (Medeiros, 2025) (EC, nd.)

As a reaction to the fast urbanization trend and economic expansion, especially in developing countries, the approach of self-sufficient cities is gaining territory as a new way to produce a balance between development and sustainability (Bolay, 2006; Losch et al., 2012; Kong et al., 2022). Considering that actually cities are open complex systems, very much dependent on the relationships with other regions in terms of import and export, any situation of natural, social, or political change that breaks these relations may cause serious problems. In this aspect the concept of a self-sufficient city and its symbiosis with sustainability is of great importance (Small, et al., 2019).

What is a self-sufficient city? According to Medeiros, "The concept of self-sufficiency entails the notions of a sustainable, circular, and carbon-neutral city" (Medeiros, 2025) Looking from an individual point of view, self-sufficiency can be defined as a situation where someone can fulfill his own needs without any external help (Ali et al., 2012; Rollock & Klaus, 2022). But the topic of a self-sufficient city, in a broader sense, is also related to social aspects and community self-resilience. In this dimension, to have a self-reliant community it is very crucial to activate efficiently the mechanisms of community participation in decision-making processes (Lertthanakulvat & Sonsuphap, 2024). The issue of sustainable development should be related stronger to community development by integrating it better in the processes of social restructuring (Bansal et al., 2019; Kumpulainen & Soini, 2019; Palma, 2019). The concept of a self-reliant community is continuously becoming more present in the discourse of environmental sensitivity. It is commonly discussed that the pillar of community development is based on the "Theory of Self-reliance" (Smith, 2021; Martin-Shields & Munir-Asen, 2022; Shunglu et al.,2022). Concerning this, communal solidarity and projects or policies that stimulate common activities are vital. Particularly cooperative actions related to the sustainability of the environment, are enormously significant to have a spirit of community among citizens. (Ali et al. 2012; Rollock & Klaus, 2022). In this regard, the allocation of flexible public spaces in neighborhoods can promote common activities among residents (Lertthanakulvat & Sonsuphap, 2024).

1.2 Pillars of Self-sufficient City

The concept of a self-sufficient city and sustainable environment is usually elaborated through fields such as carbon-free economy; circular economy; renewable energy; sustainable mobility; water management; food production; and waste management (Fig. 1.1).

According to Medeiros, every civilization is based on three resources: food, water, and energy. In this line, he suggests that to reach a carbon-free economy, developments should be focused on systems that solve the intra-city food provision, forms of renewable energy, and building of clean water recycling mechanisms (Medeiros, 2025).

Sustainable food production concerns the local processing of food and its linkage to larger market chains. In addition, the decrease in food waste is also a crucial element contributing to a sustainable system (UNep, 2018).Another component of sustainable food provision is Urban agriculture which has important socioeconomic, environmental, and health benefits (Nogeire-McRae et al., 2018). While the issue of sustainable water management addresses the approach of providing water in the long

Fig. 1.1 Pillars of self-sufficient city (diagram by author)

term without damaging the environment. Important criteria here are the maintenance of the ecosystem, reduction of pollution, water protection, and the re-use of water resources (Acciona, 2023). Probably, one of the most significant components of a self-sufficient city and a sustainable environment is the concept of renewable energy. Renewable energy can simply be defined as the energy which is obtained from natural resources. United Nations identifies six main sources of renewable energy: solar, wind, hydropower, geothermal, ocean, and bioenergy (UN, nd.). Renewable energy contributes to a carbon-neutral world and impacts climate change positively. To reach these goals it is necessary to integrate a feasible model that implements the systems of renewable energy in both public and private buildings (Dahal, 2025).

Besides food, water, and energy, **transportation and waste management** are also considered as important components of self-sufficient and resilient urban territories (Trujillo & García, 2025). The transportation sector is one of the main contributors to environmental pollution. About 25% of greenhouse gas emissions in the European region are caused by transport. In this context, new strategies that introduce smart developments, and transform the demand can stimulate a new orientation toward cleaner modes of transportation that decrease the negative impacts on our society (EEA, 2025). Sustainable mobility means providing a transport system that is smart and clean (Gavanas, 2025).

Finally, one of the most comprehensive concepts used very frequently in recent times is the circular economy. "The circular economy is a model of production and consumption, which involves sharing, leasing, reusing, repairing, refurbishing and recycling existing materials and products as long as possible" (EP, 2023). It aims to expand the life cycle of products by recycling and reusing them to generate new values. This is considered as essential initiative to have a resilient environment by shifting from a conventional linear economic model to new circular solutions (EP, 2023; Bourdin, 2025). It is crucial that the new regional and city planning paradigms embrace this concept where assets are relocated by processes of recycling and reusing, stretching the sustainable dimension of human habitat (Bourdin, 2025).

1.3 The Book Structure

The book consists of a single section composed of 12 chapters in total, in which topics are organized based on the "scope" of each study and the "dimension" of the proposed case study. It opens with an introductive part (Chapter 1) by Egin Zeka and Kreshnik Merxhani, arguing the two main concepts developed through the book and their relevance in today's approach to design and planning at all scales.

Then, it presents 10 chapters focusing on relevant concepts—analyzed within real-case studies—for sustainable habitat enhancement and self-sufficient city-making. The studies presented are based on research conducted in Albania, Greece, Iran, Italy, Egypt, and Portugal, covering a 5000 km long geographic belt cutting across the Mediterranean Basin and connecting the West and the Middle East of the central axis of that Euro-Afro-Asia Liquid Supercontinent (Fig. 1.2).

Starting from Chapter 2, the authors Antonio Acierno and Ivan Pistone adopt the Medium-sized City scale as a potential model to designate a feasible concept of "Self-Sufficiency." They investigate how this city deals with various urban, environmental, socio-economic, and political challenges and how they can display sustainable practices, highlighting their potential for resilience and self-sufficiency.

Following on, Chapter 3 examines the enduring urban project for Rome's Testaccio District in serving the historic habitat enhancement by preserving the strong local urban identity, as described by Cinzia Bellone and Paolo Trevisani. Several public works programs are currently underway in this complex and dynamic area where local and super-local functions intersect with ordinary daily needs. Documenting and interpreting this path and its related criticalities are the goals in favor of similar historic sustainable habitats.

Fig. 1.2 5000 km long geographic belt included in this research (diagram by author)

Chapter 4 speaks about the potentiality of Landscape's "Cartographies" for interpreting the local human cultural habitat in an Athen's neighborhood, by Georgia Eleftheraki. This study aims is to contribute to the documentation, and assessment of Athens' urban habitats and its early twentieth-century heritage in/and around the central areas of the city, as well as the highlight of its unprotected historical and cultural habitats.

Chapter 5 brings us to pay attention to the huge phenomenon of abandoned Factories and the rise of a potential "Third Landscape," by Mariateresa Giammetti and Kreshnik Merxhani. Cities are crossing through a fast economic process transformation, lowering the secondary to focus more on the tertiary sector. Many industrial areas exhausted their role and still stand as ruins and waste of a past role and occupied soil. The issue of abandoned factories and drosscape is particularly acute in Albania due to its history. Starting from a historical, environmental and architectural analysis—mostly concerning national case studies—the research aims to demonstrate how brownfields, disused factories and industrial landscapes may get a new meaning for a contemporary urban sustainable habitat. The article also emphasizes the human health impacts of pollution and illustrates how industrial landscape revitalization projects cannot ignore the current unsustainable legacy in terms of pollution industries produced in the places.

Chapter 6 speaks about those places Tirana leftover since the city turned towards an ideal of a "Modern City" in 1991 and the rising of "Third Landscape(s)." Fabio Naselli and Mariateresa Giammetti outline a research program developed in 2019–2023 aimed to provide an analysis of Tirana's urban open space adaptations, assessing their integrated habitats' sustainability. In any of the current visions for the next Tirana, the increasing of high-rise buildings' new fabrics is going to disregard and overlap with the fragments of the historic fabrics, leading to the erasing of these original habitats. Pursuing this way, as the study set out, the most of original open spaces will be fragments of a "third landscape"—born from this rapid growth of the formal city—which asks to be managed to recover an integrated sustainable habitat.

In Chapter 7, Marsela Plyku Demaj, Skender Luarasi, and Blerim Nika stress again the crucial relationship between heritage buildings and the urban "unit" they belong to. The study aims to reconsider the significance of "context" in light of the rapid urban transformations happening in Tirana. The discussion focuses on assessing the impact of urban changes on heritage buildings, understanding their up-to-date role in today's city, and figuring out how the context can either highlight and preserve their historic values or overshadow and diminish them.

Chapter 8, by Enrico Pietrogrande, debates the idea of using urban improvement projects as a sustainable approach to reconnect the diverse functional areas of a city by analyzing the case study of the small town of Vigonza in Italy. The analysis and design methodology are based on studies of typology and urban morphology by the Italian rationalist school, aiming to find a balanced relationship between the new and the old in urban habitats.

In Chapter 9, the article by Momen Foadmarashi and Francisco Serdoura discusses how public spaces' vitality comes from their ability to "connect people" for a more sustainable multi-cultural urban habitat. The Urmia Bazaar in Iran is a prime example of such an urban habitat that has evolved while maintaining its core role. The bazaar serves as a complex city habitat fostering relationships and connections.

Chapter 10 considers how walkable streets can enhance sightseeing in the attractivity of Tourist Places, by analyzing the case of Hamadan Inner City (Iran). The authors, Solmaz Asadpour, Momen Foadmarashi, and Francisco Serdoura, aim to emphasize the significant impact of footpaths in driving cultural tourism within the inner city. The research explores the influence of well-designed footpaths on pedestrian flow, visual understanding of public space, and the general behavioral aspects of suitable urban habitats for both pedestrians (city users) and tourists.

Chapter 11 explores the long-term value of green buildings for self-sufficient urban habitats, authored by Julinda Keçi and Geron Rakipaj. It emphasizes the importance of Life Cycle Cost Analysis (LCCA) and Life Cycle Budget Development (LCBD) in assessing the economic viability and long-term sustainability of green buildings.

In conclusion of the book, Chapter 12, Edmond Manahasa and Fabio Naselli outline the significant insights gained from the book's multi-faceted narrative in an afterword. They highlight the urgent need for effective management of built environments, particularly in urban areas, due to the limitations of natural resources on our planet. Over the past two centuries, population growth and rampant consumerism have fueled the expansion of urban settlements, leading to increased resource demand and a decrease in green and natural spaces. The book's epilogue seeks to identify key lessons essential for developing more sustainable habitats and self-sufficient cities. It emphasizes the importance of incorporating innovative concepts such as local resilience, soil preservation, circular economy, and renewable energy into updated design processes. These new considerations should be integrated into the curricula of architecture and urban planning programs, serving as both goals and lessons for thinkers at all levels.

References

Acciona. (2023, April 6). *Acciona*. Retrieved May 2024, from Sustainable Water Management: Achieving Access To Water For All: https://www.acciona.com.au/updates/stories/sustainable-water-management-achieving-access-to-water-for-all

Ali, H. M., Dom, M. M., & Sahrum, M. S. (2012). Self-sufficient community through the concepts of collective living and universal housing. *Procedia-Social and Behavioral Sciences, 68*, 615–627. https://doi.org/10.1016/j.sbspro.2012.12.253

Arcos-Vargas, A., Gomez-Exposito, A., & Gutierrez-Garcia, F. (2019). Self-sufficient renewable energy supply in urban areas: Application to the city of Seville. *Sustainable Cities and Society, 46*.

Bansal, S., Garg, I., & Sharma, G. D. (2019). Social entrepreneurship as a path for social change and driver of sustainable development: A systematic review and research agenda. *Sustainability, 11*(4), Article 1091. https://doi.org/10.3390/su11041091

Bolay, J. C. (2006). Slums and urban development: Questions on society and globalisation. *The European Journal of Development Research, 18*, 284–298. https://doi.org/10.1080/09578810600709492

Bourdin, S. (2025). Promoting local autonomy through the circular economy. In E. Medeiros, & E. Medeiros (Ed.), Self-sufficiency and sustainable cities and regions, planning for sustainable, circular and carbon-neutral development, pp. 137–148).: Routledge.

Dahal, K. (2025). Intra-city renewable enregy production. In E. Medeiros, & E. Medeiros (Ed.), Self-sufficiency and sustainable cities and regions. Planning for sustainable, circular and carbon-neutral development pp. 59–76.: Routlege.

EC. (n.d.). *European Commision.* Retrieved October 2024, from The European Green Deal: https://commission.europa.eu/strategy-and-policy/priorities-2019-2024/european-green-deal_en?utm_source=chatgpt.com

EEA. (2025, February 10). *European Environment Agency.* Retrieved March 10, 2025, from Transport and mobility: https://www.eea.europa.eu/en/topics/in-depth/transport-and-mobility

EP. (2023, May 24). *European Parliament.* Retrieved March 12, 2024, from Circular economy: definition, importance and benefits: https://www.europarl.europa.eu/topics/en/article/20151201STO05603/circular-economy-definition-importance-and-benefits

EUCO. (n.d.). *European Council.* Retrieved February 15, 2025, from European Green Deal: https://www.consilium.europa.eu/en/policies/european-green-deal/#:~:text=The%20European%20Green%20Deal%20is

Gavanas, N. (2025). Intra-city clean smart and sustainable mobility. In E. Medeiros (Ed.), Self-sufficiency and sustainable cities and regions planning for sustainable, circular and carbon-neutral development (pp. 94–116).: Routledge.

Kong, L., Mu, X., Hu, G., & Zhang, Z. (2022). The application of resilience theory in urban development: A literature review. *Environmental Science and Pollution Research, 29*(33), 49651–49671. https://doi.org/10.1007/s11356-022-20891-x

Kumpulainen, K., & Soini, K. (2019). How do community development activities affect the construction of rural places? A case study from Finland. *Sociologia Ruralis, 59*(2), 294–313. https://doi.org/10.1111/soru.12234

Lertthanakulvat, S., & Sonsuphap, R. (2024). Self-sufficiency communities for economic development: Urban perspectives in an emerging country. *Journal of Governance and Regulation, 13*(2), 68–75. https://doi.org/10.22495/jgrv13i2art6

Losch, B., Fréguin-Gresh, S., & White, E. T. (2012). Structural transformation and rural change revisited: Challenges for late developing countries in a globalizing world. *World Bank.* https://doi.org/10.1596/978-0-8213-9512-7

Martin-Shields, C., & Munir-Asen, K. (2022). Do information communication technologies (ICTs) support self-reliance among urban refugees? Evidence from Kuala Lumpur and Penang, Malaysia. International Migration Review. Advance online publication. https://doi.org/10.1177/01979183221139277

Medeiros, E. (2025). *Self-Sufficiency and Sustainable Cities and Regions, Planning for Sustainable, Circular and Carbon-Neutral Development.* New York: Routledge.

Nogeire-McRae, T., Ryan, E.P., Jablonski, B.B.R., Carolan, M., Arathi, H.S., Brown,C.S., Saki, H.H., McKeen, S., Lapansky, E. & Schipanski, M.E. (2018). The role of urban agriculture in a secure, healthy, and sustainable e food system. Bioscience, 68, pp. 748–759. doi:https://doi.org/10.1093/biosci/biy071

Palma, C. A. (2019). Conflict resolution in community development: Are the benefits worth the costs? *Critical Social Policy, 39*(2), 268–288. https://doi.org/10.1177/0261018318780162

Rollock, A. E., & Klaus, D. M. (2022). Defining and characterizing self-awareness and self-sufficiency for deep space habitats. *Acta Astronautica, 198*, 366–375. https://doi.org/10.1016/j.actaastro.2022.06.002

Shunglu, R., Köpke, S., Kanoi, L., Nissanka, T. S., Withanachchi, C. R., Gamage, D. U., Dissanayake, H. R., Kibaroglu, A., Ünver, O., & Withanachchi, S. S. (2022). Barriers in participative water governance: A critical analysis of community development approaches. *Water, 14*(5), Article 762. https://doi.org/10.3390/w14050762

Small, G. E., Mcdougall, R., & Metson, G. S. (2019). Would a sustainable city be self-sufficient in food production? *International Journal of Design & Nature and Ecodynamics, 14*(3), 178–194.

Smith, R. E. (2021). Declarations of 'self-reliance': Alternative visions of dependency, citizenship and development in Vanuatu. *Oceania, 91*(2), 236–256. https://doi.org/10.1002/ocea.5309
Trujillo, M. A., & García, M. Á. (2025). Urban sustainable development via self-sufficiency planning approaches. A theoretical discussion. In Self-sufficiency and sustainable cities and regions planning for sustainable, circular and carbon-neutral development (pp. 23–35).: Routledge.
UN. (n.d.-a) February 20. *UNDP*. Retrieved 2025, from SDG Accelerator, Supporting the sustainable and green growth of business: https://www.undp.org/sdg-accelerator
UN. (n.d.-b). *United Nations*. Retrieved January 2025, from world's population: https://esa.un.org/unpd/wpp/DataQuery/
UN. (n.d.-c). *United Nations*. Retrieved February 15, 2024, from What is renewable energy?: https://www.un.org/en/climatechange/what-is-renewable-energy
UNep. (2018). *UN environment programme*. Retrieved January 2024, from Sustainable Food Production: https://www.unep.org/regions/asia-and-pacific/regional-initiatives/supporting-resource-efficiency/sustainable-food
urbanNEXT. (n.d.). *urbanNEXT Lexicon*. Retrieved February 24, 2025, from Locally Productive, Globally Connected Self-sufficient Cities: https://urbannext.net/locally-productive-globally-connected-self-sufficient-cities/

Chapter 2
Urban Regeneration Through System Planning and a Multiscalar Approach for a Self-sufficient City: The Case of Santa Maria Capua Vetere in Italy

Antonio Acierno and Ivan Pistone

Abstract The contemporary city is characterised by a high degree of complexity in its environmental and socio-economic processes. Its recent evolution has faced problems related to spatial fragmentation and functional transformations: disused industrial and military areas now form large dross landscapes, while urban degradation and illegal neighbourhoods are common in the outskirts. Urban planning should address these fragilities through the design of inclusive community spaces, taking into account environmental aspects and new social needs. The aim of this paper is to explore local regenerative approaches to create an efficient network of services and public spaces, aimed at restoring the peripheral margins of the urban core, implementing the overall ecological quality and community assets. Indeed, sustainable planning can reveal the latent potential of forgotten urban spaces, laying the foundations for the progressive development of new areas for social use to revitalise their spatial context. Multiscalarity thus becomes a possible approach in urban planning to control the structural, long-term and relevant use of territorial resources, also by means of local-scale planning to verify the coherence of choices at the municipal level. In accordance with these theoretical premises, it is proposed the case of Santa Maria Capua Vetere, a medium-sized town with an ancient cultural heritage that suffers from several socio-environmental criticalities. In order to address these challenges, the planning structure of the city has been calibrated on the definition of multiscalar actions that will be able to adapt flexibly to the new situations generated by the socio-economic dynamics, taking into account the need to protect the urban centre and the surrounding rural areas, and conceiving transformations based exclusively on urban regeneration. Recent trends in territorial management identify for Santa Maria Capua Vetere a development scenario in which the city becomes a tertiary metropolitan pole, pursuing a compact city model through the definition

A. Acierno · I. Pistone (✉)
DiARC-Department of Architecture, University of Naples 'Federico II, Naples, Italy
e-mail: ivan.pistone@unina.it

A. Acierno
e-mail: antonio.acierno@unina.it

F. Naselli et al. (eds.), *Self-Sufficient Cities and Sustainable Habitats*, The Urban Book Series, https://doi.org/10.1007/978-3-032-10337-6_2

of a provincial ecological network that develops at the local scale into a capillary pattern of open spaces of various sizes. This paper delves into Santa Maria's planning journey, highlighting its potential for resilience and functional self-sufficiency, despite ongoing developmental hurdles: the contribution describes various research activities in which the definition of a sustainable and smart approach arise through planning application.

Keywords Sustainable planning · Medium-sized city · Santa Maria Capua Vetere · Municipal planning · Ecological planning · Green infrastructure · Self-sufficiency · Urban resilience · Urban regeneration

2.1 Urban Regeneration in the Light of Sustainable Development of the Contemporary City Systems

Cities, and by extension urban areas, have always offered numerous and varied functions to the users who inhabit them, guaranteeing the possibility of living in safety and exchanging goods, while at the same time providing places for more or less controlled social aggregation. It is possible to say that such spaces are characterised by complexity and increasing dynamism, representing themselves a driver for social-environmental and physical-economic transformation (Mizzau et al., 2025). It is therefore fundamental to set urban transformation processes on the basis of the main issues that define not only the quality of the city as a place, but also the needs of the community that lives in it, thus promoting an efficient trade-off between economic competitiveness, social cohesion and the protection of environmental and cultural features: territorial governance becomes, in a nutshell, an evaluation criterion for the management of the most pressing contemporary issues, linked to the changing status quo of the urban environments (Colantonio and Dixon, 2011).

In this light, it is worth considering the theme of urban regeneration, understood as a way of redefining public and semi-public space, aimed at the coherent and efficient organisation of the variegated demands posed by users regarding environmental, economic, functional and community issues. Although widely debated in the scientific literature, it is still difficult to provide a unique definition (Roberts et al., 2016): however, it may make sense to consider it as the design and strategic capacity to emphasise and highlight the underlying value of spaces that are substantially disused or underutilised in order to reintegrate them within the fabric of the city, promoting new uses and reintroducing them into the productive-cultural cycle of the reference society, with a perspective of long-term improvement (Chen and Duan, 2022; Evans and Jones, 2008). In this regard, we identify the prominence of the spatial dimension within urban regeneration, which is necessary as a landing ground for the supposed benefits brought by policies and plans towards the population, but also as a point of absorption of the inevitable transformative impacts, positive or negative; at the same time, the co-presence of the economic, environmental and social spheres that integrate with the physical component of places, referring in particular to urban voids, is emphasised, in line with the consolidated principles of sustainability, with a view to promote adaptive reuse practices as well as the implementation of urban services (Acierno, 2015; Omar and Saeed, 2019). According to this line of thought, the economic sphere of urban regeneration concerns the possibility of creating employment and new sources of income, guaranteeing economic growth for citizens and private investors as well as for public institutions, ensuring new forms and possibilities of investment (Domenech and Borrion, 2022; AUDIS, 2008).

From the environmental point of view, the objective is the efficient management of available resources, including the increasingly scarce resource of available land, in order to limit uncontrolled urban sprawl and further forms of pollution: in this regard, the optimisation of existing infrastructures and the promotion of alternative energy sources appears relevant (Lak et al., 2021). The social aspects, which are increasingly at the centre of the contemporary urban planning debate, are connected to guarantees of a better quality of life, in relation to spaces that reduce perceived stress but also provide adequate accessibility, in terms of spatial justice: social impacts can both generate positive changes and mitigate negative effects, considering political-environmental but also cultural aspects (Cerreta and La Rocca, 2021; Clark and Wise, 2018). Connected to this, the cultural dynamics inherent to regeneration processes privilege interventions that respect the historical-documental characteristics of places while proposing innovative solutions in line with the physical, economic and social evolution of the area (Egusquiza et al., 2021). These aspects highlight the systemic complexity in urban regeneration processes, to be included within integrated approaches aimed at unravelling their main nodes (Acierno, 2024).

At the basis of regenerative and land-use planning policies, there is thus the idea that cities alone cannot independently curb the incipient conditions of degradation they face, hence the urgency of orienting planning tools to corroborate sustainable development strategies (Manganelli et al., 2020). The transformation of urban spaces and the change of paradigm in their management is a crucial precondition for the optimal use of degraded areas and the surrounding urbanised territory, in order to achieve an all-round standard of sustainability through urban regeneration and economic promotion initiatives, since these elements, urbanisation and the local economy, are two sides of the same coin and exert a mutual influence. In this vein, urban regeneration practices can bring about a positive urban transformation, applying an integrated approach that addresses urban issues and incentivises long-lasting economic-management and socio-environmental changes, reducing critical problems and expanding community benefits (dos Santos Figueiredo et al., 2022). In summary, the contemporary city is characterised by a variegated degree of complexity that appears within its urban, environmental, socio-economic, functional and political processes.

In fact, its evolution has been constantly confronted with challenges related to land management, unequal opportunities for users, environmental hazards and the need to compensate for the progressive lack of resources and land consumption. It emerges how fragility is thus an intrinsic systemic aspect of today's city that should be flexibly faced through the design of spaces and the planning and management of uses. Consequently, it seems that absolute self-sufficiency is at the very least difficult to achieve, although this issue appears to be highly interesting in the light of the great changes humanity is going through, including the recent health emergency (Sánchez Trujillo and Huete García, 2024).

In fact, the contemporaneity is confronted with an unprecedented evolution, as we have moved from a society based on flows of renewable natural resources to stocks of non-renewable resources, generating an extremely massive population growth, on the one hand intensively exploiting the resources present on the territory, and on the other exacerbating the imbalance between growing urban areas and places harassed by a transformative stagnation. Indeed, high concentrations of users within large urban centres are by no means a synonym for sustainable progress, reflecting the high consumption of energy and materials and the massive production of waste. Indeed, large cities, both in advanced and developing countries, face significant challenges, including climate change, limited availability of energy and natural resources, and the need to preserve natural systems essential for societal well-being (Day and Hall, 2016). Managing large settlements requires constant and extensive use of energy and materials, often from distant locations, connoting such urban forms as generally unsustainable, although there are significant internal variations related to factors such as individual behaviour and income levels (van Timmeren et al., 2005).

However, there are realities whose spatial-functional organisation reveals a possible scenario of sustainable regeneration. These are limited centres in which it is possible to find elements that are needed for the well-being of the citizen. Such medium-sized towns lie halfway between large cities, which tend to fragment in their outskirts, and small-scale settlements, offering a range of residential, rural-productive and tertiary services that make it possible to identify actual polyfunctional cities (Medeiros and van der Zwet, 2020).

2.2 System Planning and Multiscalar Approach for a Self-Sufficient City

Planning the complexity of today's urban tissues requires a profound knowledge of environmental dynamics and how these influence community instances, seeking to mend the crasis between spatial fragmentation and the need for functional evolutions: in this sense, the study of abandoned areas, disused productive areas, degraded places or sites whose use is no longer compatible with today's society is valuable. It is therefore a priority to direct planning efforts towards the resolution of these fragilities through an inclusive and integrated design of community spaces, existing or potential, operating on a more or less localised scale to reveal the latent potential of these forgotten areas, in line with the action of policymakers.

Approaching issues of complexity requires a systemic approach (Bertuglia and Vaio, 2019) as proposed by recent research studies on the design of public spaces, whether green or not (Ceccato et al., 2020). As suggested by Donella Meadows (2008), a leading expert in systems thinking applied to urban and territorial structures, it is necessary to first understand the relationships between the elements of a system and its 'behaviour'. A system consists of at least three components: elements, interconnections and an overall purpose. The 'elements' are the parts that make up the system and are fairly easy to identify, especially in physical terms (a street, a park, a square, etc.) in all its characteristics (dimensions, materials, lighting, furnishings, functions, etc.). Intangible elements are more articulated and sometimes difficult to identify, such as citizens' trust, fears or social cohesion. Among the elements, we also consider the different stakeholders, from residents to commuters, police forces to shopkeepers, etc.

'Interconnections' are the communication channels, the sharing of information and the efforts that different stakeholders make to improve the quality of space. They represent the relationships that hold the elements together (Meadows, 2008).

The 'purpose' of the system might be easy to identify but is not always so clearly revealed. The purpose of a system, of a community, is the outcome of the different goals of the actors involved, which may also conflict. From this point of view, the system acts as a complex mechanism, which employs the interconnections between the elements and the different strategies of the actors to pursue a general aim. However, in order to identify the purpose of a system, one has to observe how it behaves.

Furthermore, according to the systemic theory of cities (Williams, 1970), it must be understood what the 'boundaries' are, as well as the 'equilibrium' conditions and 'governance system', within which to investigate and where the interconnections stop, which are useful to understand the reasons and dynamic mechanisms at the base of urban issues. The city is a complex system, characterised by dynamic behaviour that is difficult to understand, with unforeseen events and mutual processes that are always far from equilibrium (Batty, 2013).

Based on these premises, it can be understood the need to pursue sustainable development for the application of regenerative practices, integrating solutions for the mitigation of environmental impacts with processes of equal spatial redistribution, along with effective governance practices that support planning action, aiming for a systemic transformation to improve community well-being. However, adopting a systemic approach means recognising the importance of the interconnection of the systems underpinning the city structure, which are often fragmented and poorly integrated with each other in the territorial context of reference. Urban planning decisions must foresee the reconnection of these systems in an inter-scalar manner, supporting solutions based on local potentialities and community needs, through transformative-regenerative processes linked to effective strategies over time. The methodological proposal therefore envisages the planning investigation of the current state of the different systems underlying the city, linking them together (considering energy-environmental, socio-cultural, economic-productive and mobility aspects) in order to identify opportunities for transformability. In this way, an urban regeneration model for sustainability is proposed that is both adaptive and smart, capable of holding together the strands of the urban structure from multiple points of view.

In the light of this, this paper aims to study the case study of Santa Maria Capua Vetere and explores the planning experience linked to the recent development of this middle-size city: although the process of sustainable evolution is still far from complete, the case shows considerable potential as it is progressively developing an efficient network of services that allows for an urban system that is as resilient as possible and possesses, at least in nuce, elements of functional self-sufficiency, also due to its moderate size that makes more careful land management and planning plausible.

This proposal ultimately focuses on the operational possibility of regenerating public and private areas with a high symbolic and community value in order to identify urban poles that act as catalysts for various services for the city, proposing a kind of approach that drafts a wide territorial strategy that is constantly verified and applied by local scale interventions, related by common purposes.

The contribution summarises the authors' research experiences as scientific consultants within various academic projects providing technical-scientific support for the planning of the Santa Maria Capua Vetere municipal territory, both on a municipal and on a local-operational scale. Starting in 2017, various agreements were signed between the municipality and the University of Naples 'Federico II', in which the authors are affiliated: specifically, the research activity initially focused on the drafting of the Preliminary Urban Plan, with annexed analytical sectorial studies; subsequently, the scientific support guided the planning choices related to the final drafting of the PUC (2023), with particular attention to the enhancement of the quality of the historic centre and regeneration policies. Moreover, the authors are currently involved in the drafting of different detailed operational plans (PUA) related both to the valorisation of the cultural-historic values of the city and to the design implementation of abandoned, disused or degraded areas: specifically, the contribution provided by the authors, of a scientific and design nature, is related to the drafting of a rehabilitation project for the historical centre and three regeneration projects in the western suburbs of the municipality for the disused industrial area of the former Italtel company and for two residential areas in a state of heavy decay and lacking basic services, namely Camposorbo and Cappuccini[1] (Fig. 2.1).

2.3 The Case of Santa Maria Capua Vetere: Urban Complexity as a Driver for Regeneration Towards Self-sufficiency

In this sense, the case study of Santa Maria Capua Vetere, a medium-sized city of approximately 32,000 inhabitants in the district of Caserta, nearby Naples in the South of Italy, represents a peculiar case of planning for medium-size cities,

---

[1] The PUC of Santa Maria Capua Vetere was approved in December 2023. At the following website, it is possible to examine the documents produced, including the definition of the areas of intervention of the operational plans: https://www.santamariacapuavetere.eu/

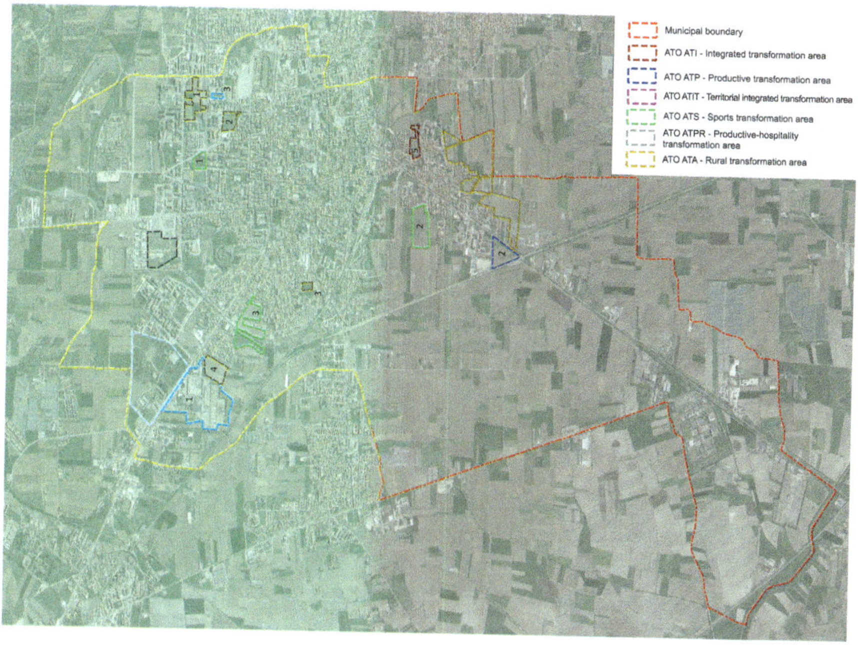

Fig. 2.1 Planning activities for the territory of Santa Maria Capua Vetere has identified not only a municipal strategy but also transformability areas in which different PUAs have been drafted: in this way, the methodological approach may guarantee an attentive control on the future urban regeneration

embedded between urban development, the rural and productive component and cultural emergencies. The municipality, with a substantially agricultural tradition, possesses in fact a dual cultural identity linked both to the evident built heritage and to the archaeological substratum of ancient matrix, currently, mostly not visible. At the same time, the forecasts of the territorial scale plans entrusted the role of future urban core to the town of Santa Maria, a strategy that has never been fully realised also as a result of the demographic shrinkage.

Despite the presence of ancient remains dating back to prehistoric ages, the original foundation of the city is rooted in the Oscan-Etruscan tradition, and in imperial times it became one of the main cities in the Roman sphere of influence, earning the title of 'altera Roma', meaning 'other Rome', an urban settlement second only to the capital of the empire (Fig. 2.2). As early as the sixteenth century, Santa Maria Capua Vetere became a privileged destination for scholars of that time, with a gradual focus on archaeological excavation activities (legitimate or clandestine) and the progressive discovery of ancient city fragments that reached their peak between the eighteenth and nineteenth centuries, with the widening of the Appian Way and the realisation of the railway. Since the second half of the twentieth century, the economic boom and the gradual expansion of the urban centre have rapidly buried the main historical-archaeological traces of ancient Capua, generating a mystification in the reading of the urban heritage, often out of sight or inaccessible due to buildings that are incoherent with the ancient remains on which they insist and with which they should instead be confronted. This phenomenon is certainly the result of municipal planning that has paid little attention to the historical fabric, minimising urban and archaeological defence and enhancement (Acierno and Lanzi, 2019).

Geographically, the municipal area lies between two adjacent metropolitan spheres (Naples and Caserta), bridging wide cultivated lands and densely urbanised zones. It is therefore no coincidence that Santa Maria Capua Vetere has played an important role in recent times, especially at the end of the last century: in the 1980s, various entrepreneurial activities flourished there, taking advantage of the municipality's strategic position, fuelling its economy on the one hand, but at the same time

Fig. 2.2 Campanian Amphitheatre is second in size only to the Colosseum in Rome. Although it is an element of incredible cultural and archaeological value, today it relates to a context in which massive urbanisation is eroding the historical substrate of Santa Maria Capua Vetere (*Source* Google earth)

changing its original structure, which gradually broke away from the agricultural tradition, in particular hemp cultivation, which was one of the main ones in the entire province until the mid-twentieth century. Testimony to this is the construction of the hemp factory complex for the processing of this product, so fundamental to Santa Maria Capua Vetere, although the production was then gradually reduced until it was replaced in the 1990s by that of tobacco, an intensive cultivation. This evolution is also linked to the growing urbanisation of the area, with the construction of sewerage works, aqueducts and new road systems that have considerably reduced the surface area to be allocated to rural activities: this is why tobacco intensification has become more profitable, also as a result of technological progress that has led to more fertile farmland and more efficient cultivation and irrigation techniques, thus increasing the output per hectare, much to the disadvantage of hemp cultivation, which is instead extensive.

The phenomenon of industrial desertification has thus led the Santa Maria production chain to turn towards new frontiers of development, in line with the development of the south of Italy: this is how the transition has taken place, from an economy based on agriculture to one with an agro-industrial imprint, which includes both important poles with a more conventional roots (such as the Parisi Mill, in the eastern edge of the municipality, the Hemp Factory and the Manifattura Tabacchi, that is the Tobacco Factory, as well as the agricultural consortium, already active in the 1970s) and others of a more innovative nature for the time, such as, the Siemens factory, located in the western area of the city and pertaining to the telecommunications sector.

The presence of these economic attractors in the municipal area resulted in the city assuming a flourishing economic character over time, also supporting local commerce, which profited from the many daily users of Santa Maria. However, to date, some of these areas are now disused and have partly been replaced by large-scale retail and companies established in the area, reducing opportunities in terms of the local supply chain. For example, the former Manifattura Tabacchi (tobacco factory), which once represented a productive hub in the area due to its direct connection with the dominant agricultural economy, has gradually lost productivity, becoming inactive and giving way to a large commercial centre, thus reflecting the shift in the local economy. A similar fate has found the agricultural consortium, which has now become a shopping centre, with little if any reminiscence of its primary function; similarly, the Parisi mill is destined to accommodate shops and supermarkets. The Siemens industrial area, after decades of strong development in which the local site housed autonomous research, design, production and marketing activities, is likewise now disused. These events show how the ending of relevant economic cycles generated within the urban system usually adjust to the economic transformation, then they may generate waste landscape when it is not possible to settle a function that would give a convenient solution for these large spatial containers (Fig. 2.3).

Alongside this is the presence, equally relevant to Santa Maria's urban structure, of actual military communities, mainly linked to the presence in the area of the massive Andolfato barracks, overlooking the large Siemens production area, around which commercial activities and services of various kinds have developed over time, fuelling the formation of new social cores: in any case, since the 1990s, the barracks

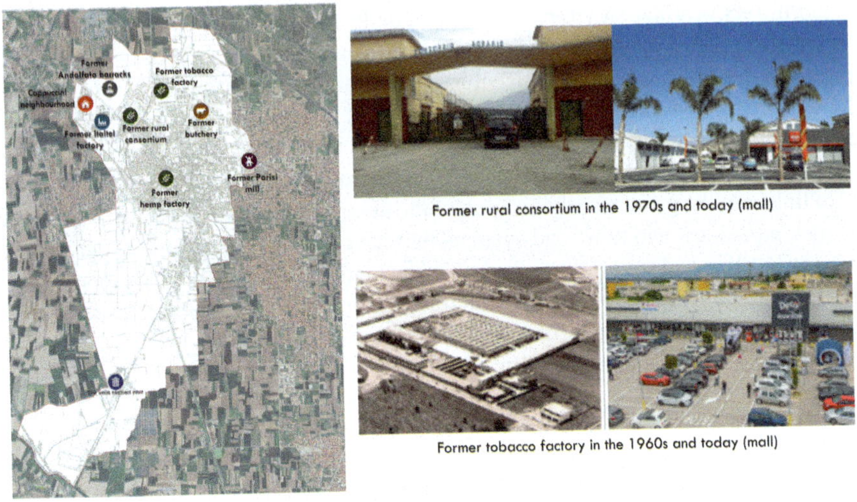

Fig. 2.3 On the left, map of the dismissed facilities and degraded areas in Santa Maria Capua Vetere. On the right, the Agricultural Consortium (above) and the Tobacco Factory (below): in both cases, the rural-productive buildings have been changed into large commercial hubs

has played a variety of roles, reflecting the changes in military needs and functions over time, hosting, among others, Civil Protection areas, warehouses and garages, as well as the site of the only military prison in Italy.

It is clear that these economic-functional evolutions, presented in an asynchronous manner in order to propose a spatial rather than chronological narrative, have contributed to create an urban pattern of great complexity that justifies the interest in the urban transformation of Santa Maria Capua Vetere. In this sense, it is valuable to emphasise the presence of 92 'neglected areas', as indicated by the Provincial Plan, i.e. open spaces not used or characterised by degradation phenomena on which a transformation can be hypothesised in order to recover them, including the former industrial and military areas for which an integrated transformability can be envisaged.

As a matter of fact, the city is characterised by the presence of potential economic-social and cultural networks, considering, among other things, the presence of polarities destined for waste disposal as well as energy production on a local scale: all these elements point towards the possibility of integrating these assets within the urban fabric, with a view to a sustainable and smart urban regeneration, aimed at a plausible degree of self-sufficiency.

2.4 Strategic Framework: The Territorial Plan and the Role of Santa Maria Capua Vetere in the Metropolitan Area

The recent history of planning in the municipality of Santa Maria Capua Vetere provides an example of administrative apathy and ineffectiveness that is widespread throughout the Campania region and common to many other medium-sized cities, especially in the Caserta and Naples areas. The planning process in this territory is mainly centred around a single main municipal town-planning instrument, the General Regulatory Plan (Piano Regolatore Generale—PRG) dating back to 1983, which has guided the transformation of nearly half a century, covering the phase of greatest expansion in response to the growing demand for residential and productive spaces. The plan, after forty years, is now to be considered exhausted in its intentions and above all completely inadequate to the needs of contemporary society (Forte, 2007).

The history of the last four decades of the 1983 PRG describes the growth on the territory of a city that in the 1950s was still called the 'city of gardens' due to the large presence of vast green courtyards, also with a rural character, that distinguished the landscape of Santa Maria. The presence of large green and rural spaces was justified by the foundation history of mediaeval Santa Maria, built around three rural boroughs whose cornerstones were the churches of S. Erasmo, S. Maria and S. Pietro (Casiello and Di Stefano, 1980). The structure of the town that had been formed until after World War II was essentially recognisable as an aggregation of three distinct districts, united only by a few main thoroughfares that still left wide green patches in the building fabric. The approval of the 1983 PRG had instead initiated the complete saturation of the rural courtyards with multi-storey buildings, as well as the inappropriate replacement of nineteenth-century and early twentieth-century buildings. Moreover, it had allowed the expansion of the city towards the outer belt by building a vast peripheral urban fabric that has rapidly suffered a fate of physical and social degradation.

The 1983 PRG was sized to reach a population of 40,000 inhabitants, with an outsized one-third increase, a threshold that was never reached and that even today, forty years later, remains stable at around 33,000 inhabitants. The PRG was implemented with widespread recourse to implementation plans in the peripheral areas of the territory by adopting multi-storey buildings. The social housing neighbourhoods present an acceptable building density and are characterised by ample space for equipped green areas and urban standards, but today they are in a degraded condition. In other cases, suburban districts were built by means of parcel plans, which were never completed and lack primary and secondary urbanisation works, and now lie in a state of neglect with empty, undeveloped lots. In many cases, spaces for public facilities were never handed over to the municipal administration, which decided in return to not maintain them (Fig. 2.4).

The history of the city's building growth is marked not only by residential expansion, but also by industrial expansion with the birth of several large industrial facilities

Fig. 2.4 Santa Maria Capua Vetere represents a type of city that is in metropolitan areas, in between rural and urban zones, in between past and present (*source* Google earth)

such as the ex-Italtel, now totally disused, and the Manifattura Tabacchi, demolished and now partially replaced by commercial warehouses. The industrial areas envisaged by the PRG were characterised not only by large facilities but also by areas of limited surface that provided for the establishment of small craft businesses. Even these areas today show degradation and settlement promiscuity due to the construction of mixed residential-industrial buildings with the obvious lack of adequate urban standards.

To summarise, the city's urban growth has largely followed the oversized forecasts of the PRG but has also been marked by unauthorised allotments in agricultural areas, industrial expansions of large complexes and small and medium-sized companies with mixed settlement. The demolition of valuable buildings in the historic centre and their consequent replacement with low-quality, invasive construction, together with the industrial decommissioning of the 1990s, complete the picture of the critical issues affecting the city of Santa Maria Capua Vetere today. The new environmental demands, territorial resilience, the fight against climate change, urban regeneration, the redevelopment of the historic fabric, the enhancement of the archaeological and cultural heritage, new forms of sustainable mobility and, finally, the need to activate social participation processes in the construction of collective goods have imposed the revision of an obsolete planning instrument, not only from a regulatory point of view, given the existence of a new regional town-planning law, but above all in order to be able to provide answers to the compelling socio-economic needs (Fig. 2.5).

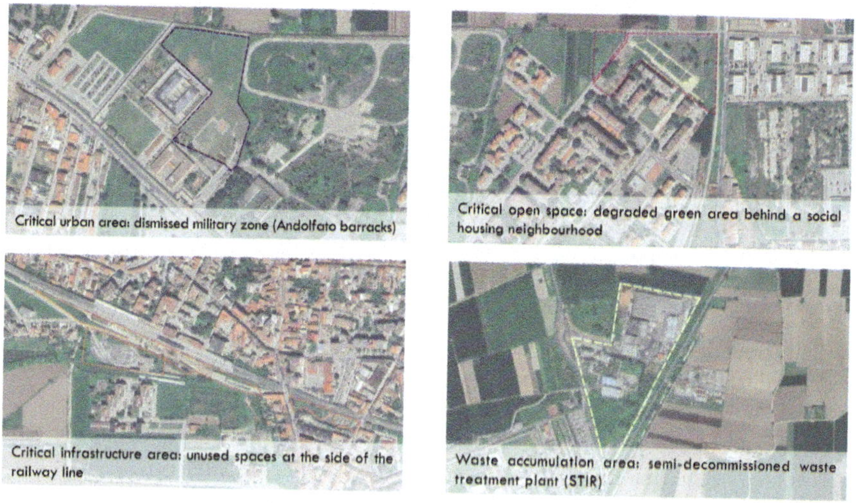

Fig. 2.5 Some of the main critical spots in Santa Maria Capua Vetere: dismissed facilities and degraded residential areas represent a pivotal topic for the enhancement of this complex urban system

2.5 Applying Multiscalarity: Local Planning Process to Enhance Santa Maria's Potential

The start of the drafting process of the Municipal Urban Plan (Piano Urbanistico Comunale—PUC), a new tool for governing the local territory, began in 2017 and was marked by the approval of the municipal administration's guidelines, through which the main objectives of the PUC and the new role of the city of Santa Maria Capua Vetere were identified, based on four strategic cornerstones: liveability, mobility/accessibility, sustainable urban regeneration and energy sustainability. From these fundamental strategic guidelines were derived further general objectives such as: the smart city, urban densification, the reduction of land consumption, the adoption of an alternative sustainable mobility (bicycle/pedestrian, tramway, etc.) and finally, the implementation of a green space network as a structuring element of the urban, peri-urban and rural landscape.

The general objectives of the plan have guided the main lines of the project structuring the future scenario of the city, which can be synthetically qualified as a 'city of nature and culture', a city made up of monuments, cultural heritage, parks, gardens, cycle-pedestrian paths and railway lines, social housing, residential fabrics recovered and integrated by services and public facilities aiming at the construction of a city resilient to the environmental and social impacts of contemporary life.

The relevant methodological model for the ecological approach of this idea of a city is represented by the concept of 'urban green infrastructure', which represents an appropriate challenge for an area that makes the nature-culture combination the backbone of its identity (Acierno, 2019).

To summarise, the objectives underlying the idea of green infrastructure, based on the nature/culture dialogue, and the redevelopment of the grey infrastructure, as well as the strengthening of the commercial/tertiary network, guided the drafting of the municipal urban plan, finally approved in December 2023.

The strategies underpinning the Santa Maria Capua Vetere PUC proposal consider first of all the municipality's role in the province zone, which certainly constitutes a territorial stronghold, not only in the Caserta area, but also as the extreme western offshoot of the Neapolitan metropolitan area.

The municipal territory is in fact located on the metropolitan urbanised boundary beyond which the vast agricultural area within the Campania plain begins.

This boundary condition reveals the dual soul of the municipality, which, on the one hand, represents a directional centre of the Caserta area due to the presence of significant higher functions (university, courthouse, hospital, etc.) and, on the other, a settlement that bases part of its economy on agriculture thanks to a very large percentage of the municipal surface area covered by fair-quality soils.

Certainly, out of the two souls, the directional and metropolitan one assumes greater prominence, as defined also by the provincial and regional plans. In particular, the PTCP (Territorial Plan of Provincial Coordination) of Caserta assigns Santa Maria Capua Vetere a metropolitan role within the Caserta conurbation, emphasising its demographic weight as provided for by the plan forecast, with the maximum allocation of 1682 housing units. The municipality, together with the provincial capital of Caserta, constitutes a strong metropolitan attraction that, in the PTCP assessments, is expected to grow in inhabitants and functions.

The provincial plan assigns the city of Santa Maria Capua Vetere a role as a metropolitan and residential attractor as well as a service provider and, at the same time, provides for the protection of the peri-urban and agricultural areas close to the urban centre. The latter is located almost entirely, with the exception of the recent suburbs, on the area on which ancient Capua stood (Pagano and Tomeo, 2021), a condition from which the need for protection and enhancement of the rich historical and archaeological heritage can be deduced.

Given the need to protect the urban centre and the nearby agricultural areas, it is consequential to hypothesise a transformation and development of the city based exclusively on urban regeneration. In line with recent urban planning orientations, declared in European and international congresses, the development scenario envisaged for the city is therefore that of a metropolitan tertiary pole, to be developed on a compact city model through urban regeneration and the definition of a provincial ecological network that penetrates right into the urban fabric. Besides, as explained before, the PTCP also identifies 'denied areas', on which transformation can be hypothesised as a priority.

The strategies for the transport system on a provincial, metropolitan and regional level confirm the trend towards the consolidation of the tertiary and residential role, envisaging the strengthening of public rail transport.

Starting from these premises of provincial planning and the declared objectives of the administration that move within a coherent logic, an articulated strategy of prefiguration of the city's future scenario is defined based on the protection and enhancement of the historic centre and the archaeological heritage, the easing of traffic flows within the urban center and the provision of a dense network of car parks and interchange nodes together with the enhancement of bicycle and pedestrian mobility, the consolidation of urban centralities by integrating services and public facilities in order to develop a multi-nodal urban framework, urban regeneration of neglected areas and disused buildings, reconnection of urban fringes through the provision of new roads and services with limited new urbanised facilities, construction of an urban green infrastructure that spreads across the territory, occupying free interstitial spaces and creating continuity within the urbanised area until it connects with external rural spaces (Austin, 2014) (Fig. 2.6).

The research supporting the municipal planning of Santa Maria Capua Vetere has thus led to the organisation of the urban fabric into three main macro-systems, within which the strategies that have resulted in the final drafting of the PUC were then channelled and that are guiding the operational projects on a local scale currently in progress. The natural-environmental and rural system is made up of the main natural and rural areas with a limited anthropic presence: in particular, it includes the areas that are exclusively or largely used for agricultural purposes; there are also areas where extensive developed agriculture is practised and those with intensive or specialised agriculture; finally, areas of high value for agricultural production, even potential, are included in this system due to their peculiar soil and climatic characteristics, steepness and terrain, as well as the presence of important agricultural infrastructures and territorial systems.

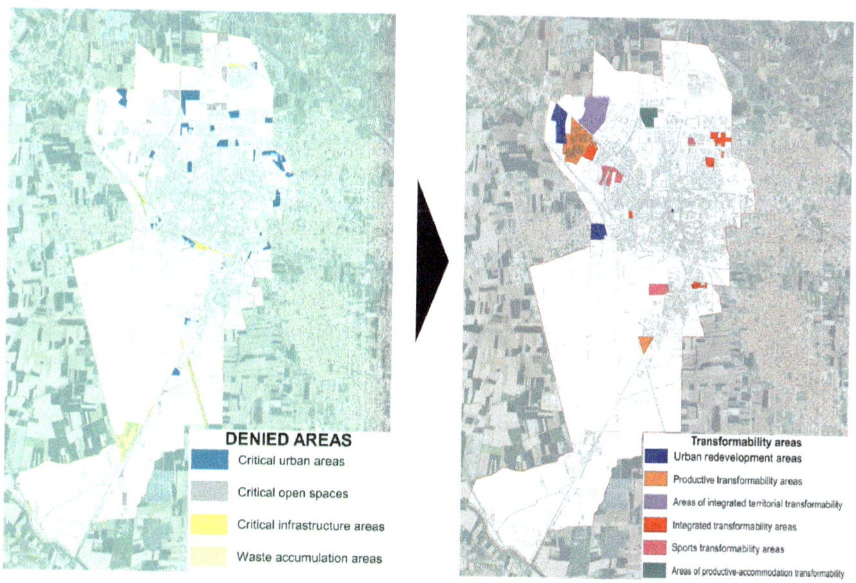

Fig. 2.6 Neglected (or denied) areas, together with the main degraded spots of the city has become the basis to identify the transformability areas for Santa Maria Capua Vetere's regeneration

The urban system is composed by the more compact and consolidated parts of the city in which human functions and activities are mostly concentrated: this system includes the urbanised areas of the consolidated or newly-formed city, the historical fabric that includes the memory and historical-artistic identity of the municipal territory, public and private open spaces, as well as buildings and special complexes of cultural, architectural and monumental interest, whether isolated or included in the ancient, civil or ecclesiastical heritage, including those connected to industrial and agricultural production; the list also covers recently built portions of the city to be regenerated, unfinished urban parts or in the process of being completed, mainly subject to implementation plans and projects in progress, and the transformation areas where the new public city is to be established. Lastly, the infrastructural-relational system is formed by the underground service and transport networks spread throughout the territory that connect the other systems, through road and rail infrastructures that are stratified at various levels throughout the territory.

2.6 Sustainable Systemic Approach for Self-sufficient Development

The project axis that deals with the historic center and the enhancement of archaeological and cultural resources must be considered the backbone of the entire development of the city. The awareness of the historical and cultural values held by the city makes it necessary to pass them on through an action of recovery and enhancement. Especially for a city that has stratified on the ruins of one of the greatest cities of Roman antiquity, according to Cicero 'the altera Roma' (the other Rome), which is the ancient Capua, the definition of a strategy for the conservation and valorisation of the existing historical, archaeological, architectural and cultural heritage has been chosen (Gigli, 2021). The 1983 PRG protected an area of just 24 hectares, limiting the definition of the historic centre to the main axis of Garibaldi Boulevard, with a few important assets such as the Campanian Museum, the Pica Barracks and the Bourbon Prison. This delimitation, implemented immediately after the 1980 earthquake, was influenced by the demands of rebuilding and therefore resulted in a reckless operation that led, in the following decades, to the continuous and systematic demolition of buildings of historical value as well as the construction of multi-storey buildings up to 22 m high, irreversibly altering the relationship with historical architecture and saturating rural courtyards and private gardens. The recognition of the exceptional values of the historic center and the construction of a city project that bases its identity on the memory of an ancient past that today becomes an opportunity for the development of new commercial, tourist and advanced tertiary activities, constitutes the main strategic guideline of the PUC. Through an in-depth analysis of the historical maps and verification with direct inspections, a new perimeter of the historic center was defined, slightly modifying the one proposed by the PTCP.

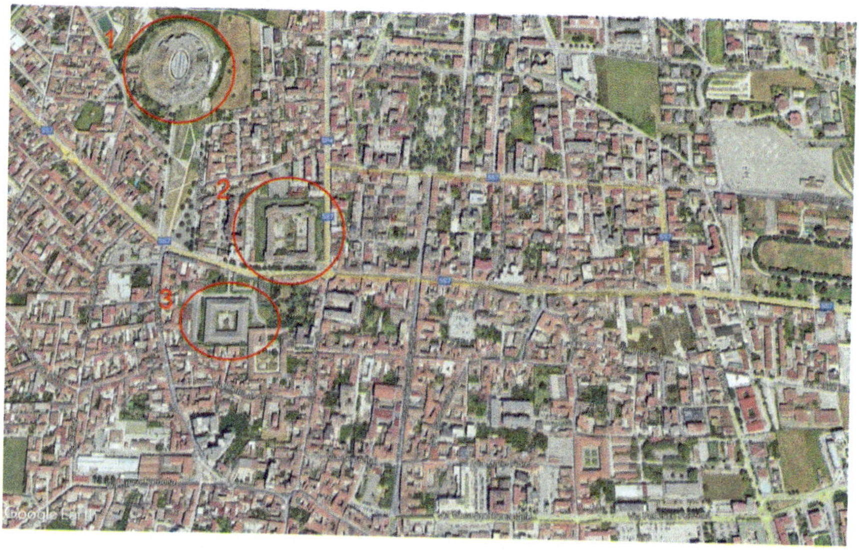

Fig. 2.7 Aerial view of the historical centre of Santa Maria Capua Vetere. It is possible to notice the main axis of Garibaldi Boulevard, the Campanian Amphitheatre (1), the Bourbon Prison (2) and the Pica Barracks, annexed to the Campania Museum (3). All these elements have been included in the wider area of valorisation of the intervention for the recovery of the historic centre (*Source* Google earth)

The strategy for the recovery and enhancement of the historic center aims to avoid a repetition of the upheavals that occurred during the 1983 PRG and, above all, to mend the existing fabric by promoting the restoration and redevelopment of buildings while respecting heights, alignments, traditional materials and, above all, the redevelopment of public spaces. In order to pursue these objectives, an Urban Implementation Plan (PUA) is being drafted, in the form of a Recovery Plan, which can provide detailed indications on each building. Therefore, this plan, which is still in progress, provides for different intervention methods depending on the degree of historical-documentary interest, the overall level of alteration, the degree of maintenance quality, and the presence of more or less concealed ancient remains. Through the drafting of a database codifying these characteristics and the definition of design guidelines for intervention, a repository of information has been created converging in further studies related to the digital enhancement of the city's archaeological-cultural heritage (Acierno and Pagliano, 2024).[2] The work on the valorisation of the city's historical heritage is also linked to the site of the Via Appia, along which the centre of Santa Maria is embedded, recently included in the UNESCO heritage list[3] (Fig. 2.7).

[2] These studies are part of the application for the candidature, for which Antonio Acierno is scientific advisor, of the municipality of Santa Maria Capua Vetere as Italian Capital of Culture 2027, an initiative proposed by the Ministry of Culture.

[3] For further information, please visit https://whc.unesco.org/en/list/1708/

Mobility is another of the key strategic themes for the city of Santa Maria Capua Vetere, which takes into account current and planned territorial initiatives regarding rail and road networks, integrated in a vision focused on the municipal area. Territorial rail connections are guaranteed by the presence of the railway network of national and interregional relevance. Instead, in order to reduce the flows within the urban centre, a ring-road circuit has been defined to bypass the urban centre and then allow traffic to enter the urban fabric at various points, acting as 'urban gateways'. This urban ring road is composed of existing road sections and a few additional new ones needed to close the circuit. The strategic mobility project proposes a hierarchical organisation of routes that intercepts flows from outside by diverting them onto the aforementioned ring road, allowing access to the different neighbourhoods through different points in the urban grid. In addition, a ring-cycle path is hypothesised with some penetrations towards the urban centrality areas consisting of the main local public services and facilities. A capillary system of public car parking spaces distributed in the peripheral urban access belt as well as in the vicinity of the main services integrates the already existing ones. In particular, parking spaces are planned in the suburbs and in the few spaces still available in the consolidated town center. The organisation of mobility within the urban center is resolved with a peripheral link that connects the city's access gates with the main axes crossing the territory and from there towards the primary transport poles such as the motorway tollgate, railway stations and the nearby airport.

From the transformative point of view, the definition of transformability areas within the municipal territory constitutes the PUC's major focus. If the categories of conservation and valorisation were mainly implemented in the strategic projects reserved for the historic center and the centralities, the one of transformability, in its various meanings contemplating building recovery, the redevelopment of degraded areas and the urban regeneration of disused areas, constitutes the content of this strategic approach. The principle of reducing land consumption and the search for higher urban quality have inspired the strategy of recovering recent settlements, which are in a state of widespread decay, and regenerating disused areas within the urban fabric (Forte, 2016). The integrated transformability areas are located in vacant sites containing 'neglected areas' in need of urban reconnection. These interventions aim to define an appropriate combination of public green areas (prevalent over the others), the slow cycle/pedestrian pathway, the presence of services and a reduced part destined for housing and commercial buildings. The spatial configurations are supposed to guarantee the continuity of the urban green infrastructure connecting the external rural areas with the green spaces inside the urban center. In the central area of the city, an increase in urban services and public space is envisaged, while on the edge of the city are concentrated transformability areas. As said before, the latter occupy neglected areas and abandoned or underutilised spaces, for which integrated transformability is envisaged, i.e. mixed residential/tertiary functional uses with areas reserved for greenery and public spaces. For each of these areas, detailed framework projects have been elaborated as part of the local implementation plan: indeed, multiscalarity constitutes a method of intervention in urban planning that controls the structural, long-term and relevant use of territorial resources, also by means of planning at the neighbourhood scale in order to verify the validity of choices at the municipal level. For this reason, the PUC of Santa Maria Capua Vetere has been guided by the multiscalar approach for controlling urban transformations. The micro-scale projects developed in the operational plans will be able to accompany and adapt flexibly to the renewed instances produced by socio-economic dynamics.

Among the areas to be earmarked for regeneration, it is worth mentioning the Cappuccini area (on the border with the municipality of Capua) that is characterised by hybrid industrial-housing functions, with warehouses and residential dwellings on the upper floors, but also the aforementioned disused former Italtel area, annexed to the overlooking and semi-abandoned Andolfato barracks, and the Campo Sorbo area, a peripheral residential district with a mainly illegal status, lacking the main community services. These areas require separate and specific regeneration and redevelopment projects. The regeneration of the former Italtel area has been addressed with the preparation of an urban regeneration plan on private initiative, which envisages an integrated transformation with functional commercial and tertiary uses and large green public spaces. The project also takes into account the possible transformation of the nearby Andolfato barracks, defining a public and semi-public space in which social accessibility is enhanced through the creation of urban public areas that connect to a new central axis. Attention is therefore paid both to the strictly technological aspects, to ensure quality in the facilities to be built, and to the merely functional sphere, to take into account the main needs of potential future users, guaranteeing equity of use and promoting socialisation and community aggregation (Fig. 2.8).

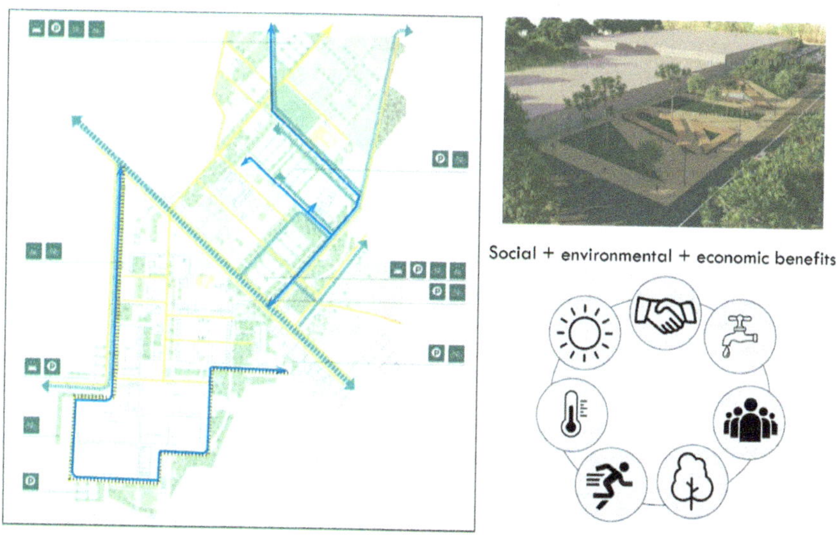

Fig. 2.8 Design vision for the areas of the former Italtel zone and the Andolfato barracks: this kind of urban connection promotes the breaking down of the functional barrier, implementing soft mobility and returning parts of the city to society

The plan, which is currently being drawn up, aims to achieve these objectives by using a transversal connecting cycle-pedestrian axis that provides access to all the spaces in the intervention site, while also identifying new routes that also serve the project's productive and commercial areas, promoting the circular economy, sustainable and zero-kilometre production and sustainable mobility practices (identified in specific mobility hubs). In addition, the intention is to achieve spatial justice with functional adaptation solutions—also respecting the principles of green planning and climate neutrality—to foster socialisation and inclusion (Fig. 2.9). Underlying these design ideas is the criterion of an open regeneration, both in terms of uses and public-private stakeholder involvement, working to truly catalyse urban renewal, creating a more vibrant and dynamic environment that can attract investment and promote long-term sustainable development. From an operational point of view, the intervention has in fact promoted the creation of an urban park, integrated with commercial and productive functions, already inherent to the area of reference: specifically, some spaces of the pre-existing buildings have been refunctionalised, implementing the quality of the walkable surfaces that give access to them. These functions are integrated with the municipal intentions of transforming this urban pole into a functional hub at territorial level.

For the southern area of the municipality, in which the Rione Sant'Andrea is located, it is planned to realise part of the ring-road axis provided for the urban centre and a secondary network of roads aimed at reorganising local traffic flows. The current centre of the Rione Sant'Andrea district lacks public facilities and has indistinct city-countryside margins with a gradual decrease in building density. Moreover, the northern part of the neighbourhood includes the decommissioned railway station site. The intervention strategy assumes to reconfigure the urban morphology especially in the urban margins with mixed residential-agricultural uses as well as through the regeneration of brownfield areas. Diffuse spaces are identified for public services in the central core in relation to the urban edges. In addition, in the area between the southernmost fringe of the built-up area and the future ring-road axis, it is proposed to create an agricultural park with annexed services in order to support farming activities. The motivation behind this planning choice is connected to the need to enhance the rural side of the municipal territory. The green areas will have to include measures to support urban drainage and the principles of hydraulic invariance, acting as a compensation ecosystem. The main services will be gathered in specific hubs for the trade of local rural products and holiday farms with related parking areas, in line with the green parking concept in order to guarantee the highest possible soil permeability. The new buildings will have to be inspired by the principles of bio-architecture in order to meet high performance requirements and to demand low energy consumption for achieving the goal of energy self-sufficiency (Fig. 2.10).

The urban and functional quality of the city is also pursued through the provision of services and equipment at different territorial levels, from the supra-municipal scale to the municipal and neighbourhood scale. The analysis of existing services as well as the forecast of needs on the basis of demographic projections revealed a prevailing shortage of equipped green spaces and educational facilities, as well as for parking, while equipment of common interest was found to be sufficient.

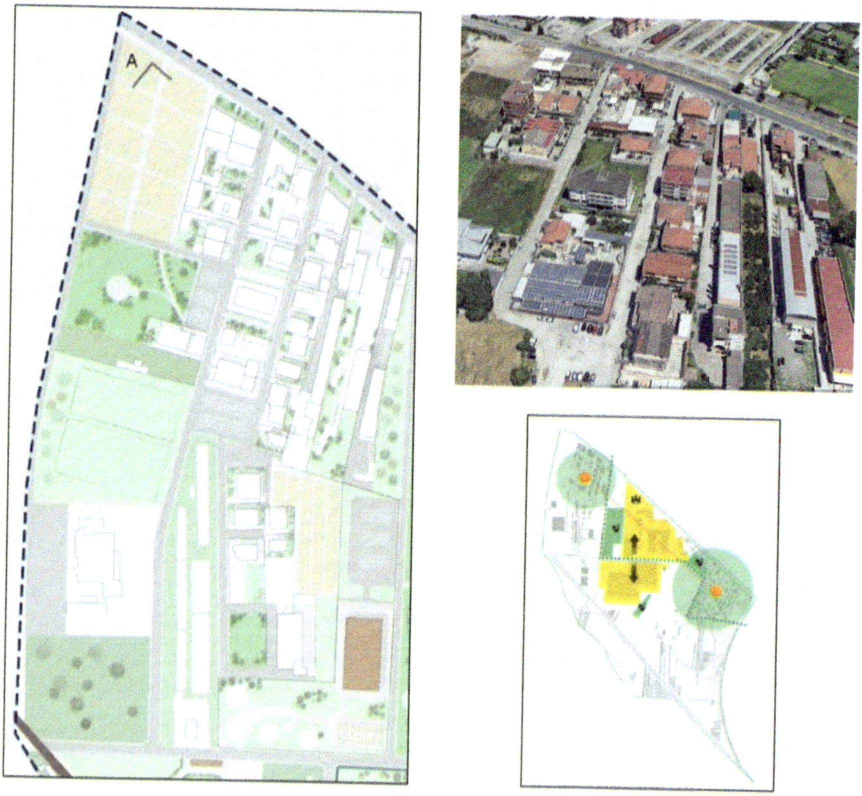

Fig. 2.9 Planning masterplan for the Cappuccini neighbourhood, nearby the former Italtel zone, proposing the enhancement of socio functional values of the site through the connection with adjacent areas and the implementation of public spaces

The arrangement of equipment areas took into account not only quantitative aspects but also positional ones, sometimes highlighting concentration in certain areas of the city. With regard to the historic center, for example, the endowment of public services and facilities of urban and territorial character is currently sufficient in some cases while it is insufficient in others, so the development strategies have taken into account the availability of disused public buildings and open spaces that are still vacant in order to assign them functions of public use aimed at raising urban quality; similarly, mobility in the historic center, which is currently congested, will be alleviated, from the management point of view, with the provision of limited traffic zones that will restrict traffic at certain times of the day and with the provision of parking spaces spread throughout the area. In this regard, municipal car parks, although quantitatively existing, are lacking in some urban areas, hence the need for their widespread distribution. Therefore, the strategy aims at the creation of new urban polarities by integrating existing services and envisaging new ones in peripheral districts.

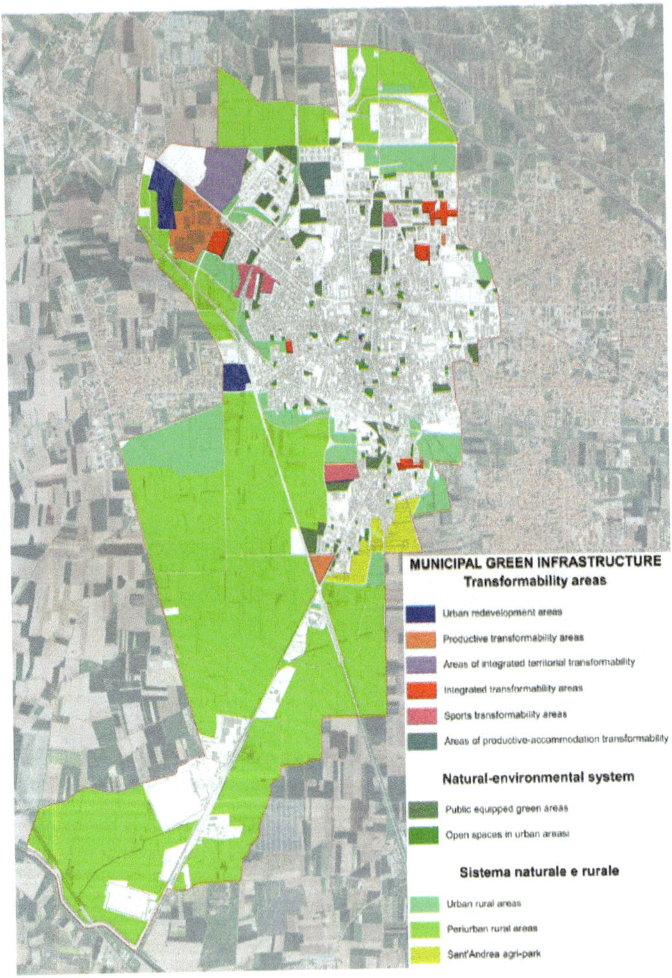

Fig. 2.10 Scheme of the integrated municipal green infrastructure: peri-urban rural areas compose its backbone; public green spaces try to cope with the issues of a very dense inner fabric; finally the transformability areas and regenerated denied areas act on a local scale

Finally, the strategic project referring to open spaces aims at the building of an urban green infrastructure that can stitch together the current chaotic mosaic of green areas, landlocked agricultural areas, neglected sites and uncultivated land into a single continuous ecological network. New areas are envisaged for public parks and equipped green spaces, as well as a circuit of cycle paths enveloping the entire city center, as mentioned above. The green infrastructure is composed not only of the open spaces and slow mobility system created by the cycle-pedestrian infrastructure, but also of the network of main open public facilities and, finally, of the large archaeological parks in the area. The PUC has set up a widespread and extensive network of green areas throughout the municipal territory, providing operational indications in the technical regulations for the planting of medium-trunked trees and hedges on individual residential lots, as well as laying down rules for the creation of gardens and urban parks, both equipped and unequipped. The municipal planning, therefore, foresees a high proportion of green spaces that will characterise the urban spaces of Santa Maria: this will sum up with the integration with the extensive endowment of cycle paths that actually run tangentially to the city, connecting with the outer countryside by enhancing the ancient Roman Centuriation axes (historical land subdivision). In this way, there will be possible, in the long term, to realise a constantly expanding urban green infrastructure that can finally link to the wider natural and rural environment surrounding Santa Maria Capua Vetere.

2.7 Conclusion: Strategic Axes of Functional Evolution and Defence of Identity Values Combined in the New Form of the Urban Plan

As this paper points out, a systemic approach makes it possible to focus on specific segments of the contemporary city, somehow enhancing its capacity to achieve a self-sufficiency that is, if not perfectly achievable, at least plausible. In fact, as mentioned above, such a criterion is difficult to fulfil, as it requires first of all adequate urban form and size to support the proper functioning of flows (energy, resources and so on), and secondly, it is essential that specific infrastructures are in place to support the community. In fact, large cities are more likely to possess most of the elements necessary for self—sufficiency, yet very often large dimensions can be a constraint, as it is difficult to administer such complex organisms where dense urbanisation often creates actual areas of marginalisation where the efficiency of flows is easily compromised.

For this reason, an urban system of limited dimensions was explored, allowing the theme of self-sufficiency to be investigated, although not offering a definitive solution, but rather analysing the characteristics necessary for the city's development to be truly sustainable and smart, based on the presence of certain assets (cultural, environmental, functional and energy) that support these concepts from a design point of view.

The PUC of Santa Maria Capua Vetere has methodologically adhered to the prescriptions of the regional law, building the planning process on the founding strategies of the project for the future city, which have been translated into a very careful and directly operational zoning for the consolidated fabric and which essentially does not require capillary and constant maintenance, while for the transformability areas it has proceeded by means of detailed prescriptions and guidelines in operational plans.

The urban planning analysis identified 'three cities' that were managed differently with respect to the values to be protected and the resources to be enhanced through specific technical tools. The 'Historic Centre', which requires capillary and specific attention for each building unit, has been regulated on a transitional basis with a few provisions referring to an implementation plan for recovery that is already being drawn up in parallel with the PUC (Acierno, Pistone et al., 2023).

The 'Intermediate City', dating back to recent decades and characterised by low-quality buildings, has been regulated by implementing technical rules for the urban control above all the renovation, demolition and reconstruction as well as the exceptional maintenance of individual dwellings. For this part of the city, it was decided to proceed with a structural plan regulation governing building transformations without resorting to the operational plan, which is inadequate for the consolidated fabric, and which instead was appropriately prepared to manage the urban transformation of disused and underused areas in the suburbs.

Urban regeneration and controlled expansion in underused brownfields are the two strategies implemented in the third city, the 'Urban Periphery', which has been valorised through the reconfiguration of disused buildings and undeveloped lots.

In the three cities, attention has been paid to the redevelopment of green interstitial spaces in order to build the urban green infrastructure by adopting incentive principles in the implementation rules, also favouring greening and the creation of permeable surfaces with economical and functional benefits (Acierno, 2021).

In short, the valorisation of the differentiated resources of the territory, from the archaeological and cultural richness of the historic centre to the heritage to be redeveloped in the intermediate city to the disused and underused areas of the periphery, has been implemented through a balanced management of the technical tools made available by the regional urban planning law, from the urban level plan to the operational and implementation plan with the application of fair land subdivision and urban equalisation. Therefore, the planning experience in Santa Maria Capua Vetere proved to be of great scientific significance, as it enabled the definition of a manageable, albeit not definitive, and replicable operational dimension for a responsible and careful development planning approach, aimed at ensuring the implementation of the flows underlying self-sufficient evolution (Fig. 2.11).

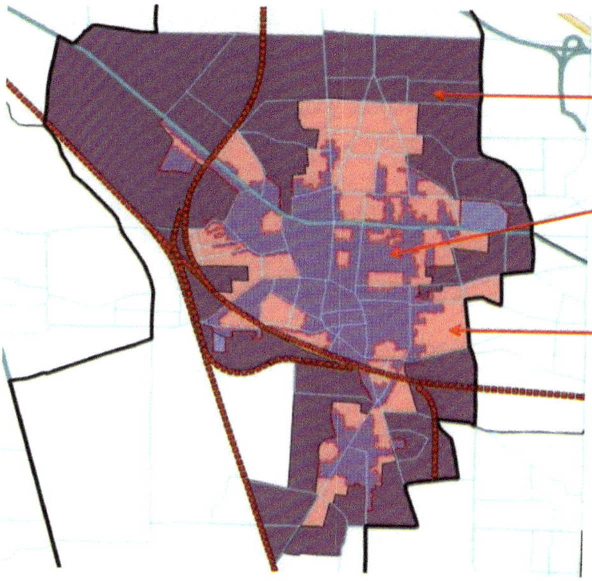

Fig. 2.11 Concept of the 'three cities' underlying the structure of Santa Maria Capua Vetere

Acknowledgement The paper and its structure have been jointly conceived by both authors. In particular, A. Acierno has written paragraphs 3 and 5, while I. Pistone has written paragraphs 4 and 6. Paragraphs 1, 2 and 7 have been drafted by both authors. All images have been produced by the authors, where it is not stated otherwise. The authors' research activity is connected to their role as scientific consultants for the LUPT research centre and the CIRURB research centre of the University of Naples Federico II in the drafting of the PUC of the municipality of Santa Maria Capua Vetere and related PUAs. Specifically, reference is made to the following research projects: 'High-profile technical-scientific support activities for the elaboration of the Preliminary Urban Plan and the Municipal Urban Plan (PUC)' (active from 2017 to 2022, for an amount of € 189.000.00); 'Building quality in the consolidated historic centre, in relation to the valorisation and regeneration policies within the regulatory and structural framework of the PUC' (active from 2019 to 2020, for an amount of € 24. 000.00); 'High-profile technical-scientific support activities for the elaboration of no. 3 Urban Implementation Plans (PUA) for the historic centre and degraded suburbs' (signed in 2022, for an amount of € 110,000.00); 'Drafting of an urban planning study for the regeneration of the disused former Siemens area located in Santa Maria Capua Vetere (CE)' (signed in 2024, for an amount of € 45,000.00).

References

Acierno A. (2015). Riempire i vuoti urbani con le infrastrutture verdi. In *TRIA*, *8*(1), FedOA Press, Naples, p. 193–212.

Acierno A. (2019). *Chromatic city: applying s—RGB design to contemporary space*, FedOA Press, Naples.

Acierno, A. (2021). Progettazione multi-scalare di infrastrutture verdi-blu: s-RGB (safe-Regenerative and Green Blue Infrastructure) Design. In P. La Greca, A. Sgobbo, & F. D. Moccia (Eds.), *Urban density and sustainability*. Maggioli Editore.

Acierno, A. (2024). Complexity in planning sustainable cities. In N. Baçi & E. Zeka (Eds.), *5ICAUD SELF—SUFFICIENT CITY—Proceedings book* (pp. 1–14). Department of Architecture Epoka University.

Acierno, A., & Lanzi, G. (2019). Strategie per la ricostruzione dell'identità negata dell'antica Capua. In *Aa.Vv., Atti della XXI Conferenza Nazionale SIU 2018*. Planum Publisher, Rome-Milan.

Acierno A., Pagliano A. (2024). The urban project for Santa Maria Capua Vetere, between contemporary city and digital archaeology. In A. Gospodini (Ed.), Proceedings of the international conference on changing cities VI: Spatial, design, landscape, heritage and socio-economic dimensions, University of Thessaly, Volos.

Acierno A., Pistone I. et al. (2023). Relazione del PUC di Santa Maria Capua Vetere.

Associazione Aree Urbane Dismesse (AUDIS) (2008). Carta AUDIS della rigenerazione urbana, Bologna.

Austin G. (2014). Green infrastructure for landscape planning. *Integrating human and natural systems*, Routledge, New York.

Batty M. (2013). The new science of cities. MIT press, London.

Bertuglia C., Vaio F. (2019). Il fenomeno urbano e la complessità: concezioni sociologiche, antropologiche ed economiche di un sistema complesso territoriale, Bollati Boringhieri, Torino.

Casiello S., Di Stefano A. M. (1980). S. Maria Capua Vetere: architettura e ambiente urbano, Editoriale Scientifica, Naples.

Ceccato, V., Canabarro, A., & Vasquez, L. (2020). Do green areas affect crime and safety? In V. Ceccato & M. K. Nalla (Eds.), *Crime and fear in public places*. Towards Safe, Inclusive and Sustainable Cities, Routledge.

Cerreta M., La Rocca L. (2021). Urban Regeneration Processes and Social Impact: A Literature Review to Explore the Role of Evaluation, In O. Gervasi et al., Eds., Computational science and its applications—ICCSA 2021. ICCSA 2021. Lecture Notes in Computer Science, 12954, Springer, Cham.

Chen X., Duan J. (2022). What they talk about when they talk about urban regeneration: Understanding the concept 'urban regeneration' in PRD, China, in Cities, 130, Elsevier, Amsterdam.

Clark J., Wise N. (Eds.) (2018). Urban renewal, community and participation. Theory, Policy and Practice, Springer, Cham.

Colantonio A., Dixon T. (2011). Urban regeneration and social sustainability: best practice from European cities. Wiley, Hoboken.

Day, J. W., & Hall, C. (2016). The myth of urban self sufficiency. In J. W. Day & C. Hall (Eds.), *America's Most sustainable cities and regions* (pp. 25–36). Copernicus.

Domenech T, Borrion A. (2022). Embedding circular economy principles into urban regeneration and waste management: Framework and metrics. Sustainability, 14(3), MDPI, Basel.

Egusquiza A., Zubiaga M., Gandini A., de Luca C., Tondelli S. (2021). Systemic innovation areas for heritage-led rural regeneration: A multilevel repository of best practices. Sustainability 2021, 13(9), MDPI, Basel.

Evans J., Jones P. (2008). Rethinking sustainable urban regeneration: Ambiguity, creativity, and the shared territory. Environment and Planning A: Economy and Space, 40(6), SAGE, Thousand Oaks.

Forte F. (2007). L'ideazione urbanistica. Architettura—Città. Un programma ed un progetto. S. Maria Capua Vetere (CE) nella regione Campania, Edizioni Scientifiche Italiane, Naples.

Forte F. (2016). Un futuro per le città storiche tra inquietudine, precarietà e impegno. Santa Maria Capua Vetere, Aracne editore, Rome.

Gigli S. (2021). Studio sulla valutazione dell'interesse archeologico, della relazione archeologica e della carta del rischio e delle potenzialità archeologiche a corredo del P.U.C.

Lak A., Sharifi A., Khazaei M., Aghamolaei R. (2021). Towards a framework for driving sustainable urban regeneration with ecosystem services. Land Use Policy, 111, Elsevier, Amsterdam.
Manganelli B., Tataranna S., Pontrandolfi P. (2020). A model to support the decision-making in urban regeneration. Land Use Policy, 99, Elsevier, Amsterdam.
Meadows D. H. (2008). Thinking in systems: a primer. Chelsea Green Publishing, White River Junction.
Medeiros E., van der Zwet A. (2020) Sustainable and integrated urban planning and governance in metropolitan and medium-sized cities. Sustainability 2020, 12(15), MDPI, Basel.
Mizzau L., Montanari F., Rodighiero S., Rinaldini M. (2025). Urban regeneration projects and the role of municipal officials: A strategy-as-practice perspective. European Urban and Regional Studies, SAGE, Thousand Oaks.
Omar N. A., Saeed E. H. (2019). Urban voids as potential resources for the city development. *JES. Journal of Engineering Sciences*, *47*(5), p. 585–600, Assiut University, Faculty of Engineering, Assiut.
Pagano M., Tomeo A. (2021). Capua la seconda Roma. Nuovi studi e ricerche, Belle Epoque Edizioni, Naples.
Roberts P., Granger R., Skyes H. (Eds.) (2016). Urban Regeneration, SAGE, Thousand Oaks.
Sánchez Trujillo, M. A., & Huete García, M. A. (2024). Urban sustainable development via self-sufficiency planning approaches. A theoretical discussion. In E. Medeiros (Ed.), *Self-sufficiency and sustainable cities and regions*. Planning for Sustainable.
dos Santos F. Y. D., Prim M. A., Dandolini G. A. (2022). Urban regeneration in the light of social innovation: A systematic integrative literature review. Land Use Policy, 113, Elsevier, Amsterdam.
van Timmeren A., Röling L. C, Tawil E. D. (2005). Urban and regional typologies in relation to self-sufficiency aiming strategies. In Sustainable development and planning, 1(2), WIT Press, Southampton, p. 1095–1105.
Williams H. E. (1970). General systems theory, systems analysis, and regning: An introductory bibliography. Exchange Bibliography No. 164, Council of Planning Librarians.

Chapter 3
The Urban Project of the Testaccio District in Rome: How a Compact Fabric Can and Must Accommodate Strategic Transformation Initiatives for a Sustainable Habitat Progress

Cinzia B. Bellone and Paolo Trevisani

Abstract The Testaccio district is a well-defined portion of the urban reality in the Metropolitan Area of Rome—the Capital of Italy—characterized by a compact and consolidated historical fabric. In an already highly complex area where local functions overlap with daily needs and attempts to preserve its own identity, as well as metropolitan functions linked to the presence of strong attraction centres (universities, archaeological areas, etc.), public works programs financed by the funds of the 2025 Jubilee and the PNRR (National Recovery and Resilience Plan) are underway. Additional funds are added to these, and they are invested through agreements between the Municipality of Rome, the Faculty of Architecture (Roma Tre) and the Academy of Fine Arts. Together, these projects will reshape the current structure of the area, its accessibility, and its intended uses. The proposed functions are deeply transformative (universities, tourism, environment), and the investment amounts to approximately 130 million euros. A unique dynamic is thus being created in the current landscape of this significant neighbourhood in the city of Rome. All these initiatives appear significant, and some were indeed anticipated. However, the urgency of the times and the complexity of the interventions severely test the Public Administration, which seems to focus on the boundaries of each individual project. Consequently, the investigation into any distorting effects that the sum of these transformations might impose on the territory has been neglected; an issue raised instead by city associations. This ongoing dialogue, initiated promptly, represents a precious opportunity to evaluate participation mechanisms, tested on an object of considerable complexity. Furthermore, it tests the governing capacity of the Public Administration to manage not only the indicated programs but also the transformations that

C. B. Bellone (✉)
Università Guglielmo Marconi, Rome, Italy
e-mail: c.bellone@unimarconi.it

P. Trevisani
Consulta Rione XX, Rome, Italy
e-mail: paolo.trevi@fastwebnet.it

F. Naselli et al. (eds.), *Self-Sufficient Cities and Sustainable Habitats*, The Urban Book Series, https://doi.org/10.1007/978-3-032-10337-6_3

will influence circulation, the network of commerce and services and the real-estate value of homes and rents. Finally, a fundamental point that the case study aims to demonstrate is that it is indeed possible to establish self-sufficient and sustainable aspects within the bottom-up approach, particularly when supported by effective and responsible practices from the Public Administration. This approach is not only about grassroots initiatives but also about how these initiatives can be effectively integrated and enhanced through public-sector collaboration. When local communities are empowered to lead with their knowledge, needs, and priorities, and when these efforts are met with proactive engagement, funding and regulatory support from the Public Administration, a truly sustainable and resilient urban transformation can occur.

Such an approach requires a commitment to long-term planning and flexibility, ensuring that the solutions implemented are adaptable to the evolving needs of the community. By fostering an environment where both local actors and public authorities work in tandem, the project can achieve a balance between ecological sustainability, social equity and economic viability. In this way, the public sector can play a crucial role, not only by providing financial resources but also by facilitating dialogue, ensuring that all stakeholders—from citizens to local businesses—are actively involved in the decision-making process. Thus, a truly collaborative and sustainable urban model can emerge, offering valuable lessons for other cities and communities facing similar challenges.

Therefore, documenting and interpreting the phases and results of this dialogue is the main objective of this article.

Keywords Urban project · Major events · Historic city · Participatory process

3.1 Introduction: The Relationship between 'Urban Project and Exceptional Public Intervention Programs' (PNRR and JUBILEE 2025)

Urban planning is a constantly evolving discipline, influenced by a multitude of factors, including socio-economic dynamics, population needs and exceptional events. In this context, the National Recovery and Resilience Plan (PNRR) and the 2025 Jubilee represent two important public works programs that pose significant challenges and opportunities for the design and development of cities. This article aims to explore the relationship between urban projects and exceptional events, analysing the implications of the PNRR and Jubilee 2025, and discussing strategies to address these challenges.

The PNRR, born in response to the economic and social crisis caused by the COVID-19 pandemic, aims to revive the Italian economy through targeted investments in key sectors, including infrastructure, digitalization and environmental sustainability. In this context, urban planning plays a crucial role in determining how these funds will be used to improve the quality of life in Italian cities.

One of the main challenges for urban planners is integrating the objectives of the PNRR into existing urban development strategies. This requires a holistic, long-term vision that takes into account the present and future needs of the local community, as well as environmental sustainability and resilience to climate change.

The Jubilee, an internationally significant religious event held every 25 years in Rome, represents a unique opportunity for the city to welcome millions of pilgrims and visitors from around the world. However, this exceptional event poses significant challenges in terms of mobility, security, and the preservation of cultural heritage. Urban planning plays a crucial role in preparing the city for Jubilee 2025, ensuring that infrastructure and services are able to meet the needs of such a large and diverse crowd. This requires an integrated approach involving all stakeholders, from Public Administration to local communities and the private sector.

Certainly, the interest in transformations concerns the entire city, but we will focus mainly on the Testaccio neighbourhood, on which we have already deepened our studies[1], without neglecting even the broader effects up to the entire Ostiense area, once regulated by the Ostiense-Marconi urban project, currently under revision.

The investigation we have carried out is based on theories, always current, that pervade scientific literature. The critical approach used to investigate the 'Testaccio case' is inspired by Harvey's (2008) theories on the 'right to the city', which emphasizes the socioeconomic impacts of urban transformations. Avoiding gentrification phenomena and working to preserve social cohesion become priority objectives, seeking inclusive solutions that actively involve the local community. Promoting equitable urban development should be a fundamental pillar of planning.

The legacy of Jacobs (1961) couldn't be ignored, which emphasized urban vitality and diversity, suggesting that new infrastructures should organically integrate with the existing urban environment. This approach implies urban design that fosters connection between public spaces, encourages functional diversity and promotes a sense of community belonging.

The theme of environmental sustainability, a cornerstone in the contemporary agenda, should permeate all phases of the urban planning process, inspired by Beatley's (2012) ideas on 'Green Urbanism', which offer practical approaches, pushing towards ecological solutions, the creation of urban parks, and the adoption of sustainable construction techniques. Planning should aim to mitigate environmental impact, promote urban biodiversity, and encourage sustainable transportation.

The active inclusion of the local community, according to Arnstein's (1969) principles of participation, is essential to ensure that planning responds to the real needs of the population. Participatory tools and public consultations should be adopted to encourage meaningful involvement, ensuring that planning reflects the expectations and visions of the community.

---

[1] Bellone (2021), I Prati del Popolo Romano. La partecipazione come sperimentazione per il progetto dello spazio urbano, Franco Angeli, Milano-Roma. ISBN 9788835122555, On line 8,835,122,554

Urban governance, as outlined by Pierre and Peters (2000), must be solid and inclusive. Effective governance should foster collaboration between public and private actors, facilitating cooperation and coordinating efforts to implement urban transformation projects coherently.

Finally, the analysis of specific case studies, with a critical eye on past successes and mistakes, provides valuable guidance. These cases can illustrate practical challenges and solutions adopted, informing decision-making processes and improving future urban planning. You find an interesting reconstruction of the themes in Bellone and Trevisani (2020), Bellone et al. (2018, 2019a, b, 2020).

Ultimately, the authors are convinced that urban planning in response to events like the PNRR and Jubilee requires a multifactorial approach that balances socio-economic, cultural and environmental needs. Through an inclusive, sustainable and community-oriented process, urban planning can emerge as a positive force to transform cities into more resilient places suited to the needs of contemporary society.

3.2 The Testaccio District and Planned Interventions

The Testaccio neighbourhood is a well-defined part of the urban reality of the Metropolitan Area of Rome. In particular, it falls within the boundaries of the first Municipality, where a significant portion of the historical centre of the Capital is located. It covers an area of 60 hectares and has a population of 7600 inhabitants, tightly delimited by geographical barriers (Monte Aventino, Tiber River), artificial ones (railway lines) and the archaeological layout of the Aurelian Walls. Compared to other neighbourhoods in Rome, it is of medium-small size. It is characterized by a compact and consolidated historic fabric. Until now, its robust popular identity, stemming from a history of social hardship, has effectively countered the drift of gentrification, with all the positive effects that follow. In this already complex territory, local functions overlap with daily needs and the desire to preserve the identity of the place. In addition, there are metropolitan functions linked to the presence of strong attraction centres such as universities and archaeological areas.

Today, we are witnessing a further turning point. The interventions planned by the PNRR and the Jubilee 2025 are related to a higher-level functional profile, of a metropolitan/territorial nature. Their parallel and concentrated implementation in a short time could result in a reshuffling of functions, as well as further pressure on the settlement system and circulation. Therefore, it is a matter of general interest to anticipate the outcomes of the upcoming interventions, to assess their impact, both individually and especially in overall terms, and to suggest complementary designs, or to request adjustments and modifications so that beyond all good intentions, the new arrangement to which the neighbourhood appears destined is advantageous and not regressive.

To the two programs mentioned above, an additional initiative must be added. During this same period, the Municipality of Rome is nearing the almost complete recovery of the former Slaughterhouse. Completed in 1891 as part of the modernization efforts transitioning papal Rome to the functions of the new Capital of Italy, it was one of the most advanced facilities in Europe at the time and is now set to become an internationally renowned cultural centre. It is the most significant project in the district and one of the largest under development in Rome.

The municipal initiative (integrated, as we will see, with a PNRR project) involves granting some pavilions on a free loan basis to Roma Tre University and the Academy of Fine Arts, which will restore and repurpose them using funds from the Ministry of Universities.

The overall proposals for the entire district currently being implemented are oriented towards highly transformative functions such as university education, quality tourism, culture and environmental protection. The total investment amounts to approximately 130 million euros, creating a unique dynamic in the neighbourhood. While significant, some of these initiatives were already planned. However, the urgency of the times and the complexity of the interventions severely test the Public Administration, which is focused on the details of each individual project. This has led to a neglect of the analysis of the possible cumulative impacts of the transformations on the territory, a concern raised by city associations. The ongoing dialogue offers a valuable opportunity to evaluate participation mechanisms in such a complex context. Furthermore, it tests the governance capacity of the Public Administration to address not only specific programs but also the transformations that will influence mobility, commerce, services and the real-estate value of the area.

The importance of documenting and interpreting the phases and results of this dialogue is the main purpose of this article.

3.3 Planned Interventions

The renovation of the former Slaughterhouse, as already mentioned, consists of the following interventions:

a. **PNRR. City of Arts**. Restoration, and redevelopment of pavilions. Proceeding administration: Capitoline Superintendence. Allocation: **18.71 million euros**.

b. **Municipality of Rome**. Grant in favour of the Academy of Fine Arts for the pavilions and related areas within the former Slaughterhouse, to be used for Higher Education in the Arts and Music (DAMS). The intervention is funded with **22 million euros** from the university construction funds at the Ministry of Universities.

c. **Municipality of Rome**. Grant in favour of Roma Tre University for the pavilions and related areas within the former Slaughterhouse, to be used for research activities and the establishment of university residences. The intervention is funded with **44 million euros** from the university construction funds at the Ministry of Universities.

In addition to the already mentioned City of Arts in the PNRR, we find:

d. **Linear Park of the Aurelian Walls** from Porta San Paolo to the Tiber. Restoration. Reference structure: Capitoline Superintendence. **€2.34 million**. The overall intervention consists of 6 lots for a total of **23.167 million euros**.

e. **Servian Walls and other sections**—Conservation restoration and enhancement. Reference structure: Capitoline Superintendence. Total amount: **€2.34 million** for all surviving sections.

f. **IV November School**, based on some redefinitions, enters the program with a double intervention for a total of **€8.715 million.**

On the detailed Jubilee 2025 intervention program, we find:

g. **Bridge of Industry Renovation**. Implementing body: Anas S.p.A. Total funding amount: **€7.85 million.**

h. **Opening of accesses on the Marconi side of the Roma Trastevere station.** Proposing administration/implementing body: Rete Ferroviaria Italiana (RFI). Total amount: **€8.00 million.**

i. **Tiber Park Rome South—Left Bank (Ponte Marconi—Ponte Garibaldi).** Proposing administration/implementing body: Special Superintendence of Archaeology, Fine Arts and Landscape of Rome. Total cost of the intervention: **€10.00 million**. Preliminary operations at the expense of the Region included dock cleaning for **€500,000.**

In conclusion, according to current data, the total intervention amounts to **130 million euros**, with approximately **90 million euros** inside the former Slaughterhouse and **40 million euros** outside.

The intervention in the former Slaughterhouse, as mentioned above, is by far the most significant, not only for its economic and structural dimension but also for its role as a European-level cultural attractor, whose gravitational scope largely exceeds the city of Rome and the Lazio region. In the following map, only the interventions adjacent to the former Slaughterhouse itself have been identified. The others fall outside the map, not far away (Fig. 3.1).

3.4 The PNRR on the Current Urban Development Plan

The most significant characterization that the Master Plan (P.R.G) identifies are three 'Strategic Programming Areas' out of the five indicated, which intersect in the territory under consideration. These areas, defined by natural elements, axes and historical routes interrelated with each other, have a morphological and functional structuring significance for the city of Rome, both at the urban scale and at the broader territorial scale. This circumstance creates interrelations that would warrant a targeted architectural and urban design intervention, especially considering that additional prestigious functions are now destined for this special area.

The Tiber River, the Aurelian Walls and the railway belt thus have an exceptional intersection point in Testaccio. Moreover, the Parco dei Fori, in its ongoing reconsideration, establishes a new proximity to the Emporium, the port of the imperial city located here.

The first intercepted area is the Tiber River, with all its advantages for leisure activities, walks and sports. It is not just a waterway but also an air passage channel. Together with the Aniene River and the ditch system, it acts as an air conditioner

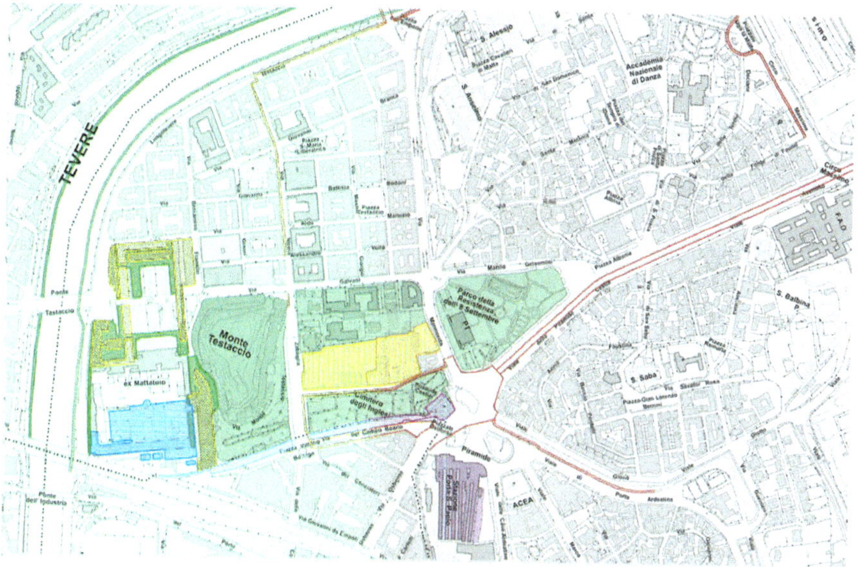

Fig. 3.1 Localization of interventions within the boundaries of the Testaccio district. *Legend of Interventions*: Linear Park of the Aurelian Walls (purple striped). City of Arts at the former Slaughterhouse (green striped). Teverever (double dashed green line). Completion of the Faculty of Architecture settlement (red striped). New structures for University-level Artistic and Musical Higher Education (light blue). *Legend of the Bike Path System*: Existing bike paths at urban level (solid red line). Planned urban-level bike paths (dashed red line). Bike path below the right bank of the Tiber River (solid green line). Planned bike path below the left bank of the Tiber River (dashed green line). GRAB (Great Ring Road for Bikes) (double marked red line). *Other Indications*: Ostia/Lido railway stop and Piramide metro station (purple). Area subject to citizen participation procedures (green and yellow areas)

for the city. Traditionally, the Tiber has also been assigned a role in favour of urban circulation. However, in the area under examination, particularly on the left bank, the river defines the edge of an area with one of the largest urbanistic and historical-monumental legacies of ancient and Renaissance Rome. Yet, it is precisely this bank that borders the Capitoline Hill, the Circus Maximus, the Aventine and the Emporium, which is now almost entirely impassable. Thus, there are no access possibilities, and consequently, broader city movement possibilities via cycle and pedestrian routes are also obstructed.

The second area is that of the Aurelian Walls, a pride of the Capital for the rarity and extent of the monument (13 km), and for the opportunities it can provide, equipped as a linear park, still for cycle and pedestrian circulation. The walls have a circular path, cutting across the consular roads, which is obvious given the defensive function they were meant to serve. However, this tangential path appears particularly useful today for movements, always cycle and pedestrian, in the 'star-shaped city'.

Therefore, rightly, the PNRR takes charge of the Walls with an overall allocation of approximately 23 million (Porta Latina-San Sebastiano; Porta San Paolo-Tiber; via

Campania-Corso d'Italia; Viale del Policlinico-British Embassy; piazza S: Giovanni-Viale Castrense; Porta Latina-Porta Metronia).

Another strategic piece of equipment is represented by the railway ring. Its role is to attract private traffic from the province and the metropolitan area towards the parking system at the railway stations and towards the interchanges of the railway with the city's public network, which now also includes cycle and pedestrian paths.

Finally, the Central Archaeological Park. This area of Rome, following the closure of Via dei Fori Imperiali to private traffic, is currently undergoing a profound reinterpretation after the archaeological excavations carried out there in recent years. The aim of the new CArMe project (Monumental Archaeological Centre) is to move beyond the twentieth-century interpretation, i.e., from the theatrical representation to the proximity of the ancient, from the view from the car windows to the multiplicity of landscapes, and ultimately to open up the largest archaeological site to urban daily life.

In this approach, the entire area of the Roman port on the Tiber, the Emporium, the spaces destined for the storage and handling of goods, in short, the entire Testaccio area, appears as the prosaic counterpart of imperial splendour. Rome did not only demand precious marbles and luxury but also wine and oil, foodstuffs, and everything necessary for the sustenance of its population. In this part of the ancient city, boats brought what was needed, and its distribution was managed here.

In conclusion, according to the intentions of the PNRR and the Jubilee, the Tiber River is destined to shift from a fracture in the settlement to a hub of new types of connections; the linear park of the Aurelian Walls is enhanced as an illustrious monument but also because it offers the opportunity for a new urban park and a soft mobility path in the heart of the ancient city; the railway station system confirms its contribution to alleviating urban and metropolitan traffic; the reinterpretation of the central archaeological area proposes to citizens and tourists a broader and shared approach to its legacy, also through its extensions that extend towards the Pyramid and towards the large area of its river port. The redefinition of all these traces and facilities redraws roles and creates new expectations in the territories crossed. Among these, also in the area of interest here.

3.4.1 Intermediate Designs: The Tiber Basin Authority Plan, the Historic City Enhancement Scopes, the Ostiense-Marconi Urban Project

The Tiber River enters the area under examination through the Strategic Plan P.S. 5 for the metropolitan stretch from Castel Giubileo to the mouth. This plan, drafted by the River Basin Authority, identifies possible actions and projects compatible with the Tiber and its banks. The Teverever intervention refers to this programmatic and project framework, which takes charge of the stretch from Ponte Garibaldi to Ponte Marconi.

Another indication affecting the area under examination comes from the 'Areas for the Enhancement of the Historic City'. The General Town Plan identifies places that over time have not achieved or have lost the defined characteristics of identity for the historic city or are characterized by the presence of buildings and structures no longer in use and convertible to new uses, or that show evident phenomena of physical and functional degradation. These characteristics are found in the area of Porta Portese and Clivo Portuense, on the right bank, beyond the Testaccio and Sublicio bridges. Here, there are opportunities for local and urban redevelopment, the recovery of abandoned buildings (e.g., the historic railway station), and a significant set of cultural assets. The area undergoing urban transformations, both ongoing and planned, has already been the subject of initial investigations by municipal offices (Simplified Preliminary Layout Scheme SAP), particularly useful for understanding the dynamics of the broader territory.

Further indication comes from the Ostiense—Marconi Urban Project. This initiative began in 1999. Its forecasts have been decisive, for example, for the establishment of Roma Tre University along Via Ostiense. The relocation of the Faculty of Architecture to the former Slaughterhouse, one of the main interventions currently underway, is due to those indications.

Many other predictions, however, have remained on paper, including the road network scheme, partly unrealized, but partly lacking in the cycle-pedestrian component, which was poorly considered at the time but is now destined to absorb increasingly significant portions of city travel.

3.5 Initial Considerations on the Effects of the Described Interventions: Mutual Constraints and Impact on the Hosting Territory

The expansion of the Architecture faculty, one of the largest construction sites in Rome at the moment, and that of the Academy of 'Belle Arti' (as anticipated, it will evolve into the DAMS) present uncertain estimates regarding student influx. The design dates back to 2003, and the initial estimates need to be reconsidered. However, we can already be certain that the former Slaughterhouse, which, as mentioned, also includes the exhibition spaces called 'The City of Arts', will fully develop its role as a cultural attractor in a context that will undoubtedly surpass the boundaries of Lazio and Italy to engage the international scene.

University activities are rightly considered high restructuring capacities, capable of upsetting the balance of large territories with the force of an immediate demand that develops in a multitude of directions. Rome has already experienced the effects of these dynamics in the neighbourhoods around the university campus.

For at least a decade, Testaccio has been undergoing development towards areas that were abandoned until recently. It was the Architecture faculty that began creating the first elements of attractiveness within the former Slaughterhouse, then it was the turn of the new market, which made the space available at Piazza Testaccio and allowed for its redevelopment. In parallel, an archaeological excavation campaign has brought prominence back to the port complex of the Emporium. The construction of the new market has precisely shed light on the functions performed by the ancient shipping offices, tasked with saving recoverable amphorae and sending those of wine and oil, which, being permeable, were irretrievable, to the landfill (Monte dei Cocci). Added to this are the two cemeteries, the Acattolico, which has always fascinated tourists, especially those from Northern Europe, and the Cemetery of Anglo-American Fallen (World War II), which also attracts Americans.

For some years, groups of tourists attracted by these attractions have begun to traverse the district. This is a higher-quality tourism compared to the intensive tourism around the Colosseum. Visitors must have already seen the main monuments, so we are dealing with individuals who have planned a long stay or have chosen Rome for a second visit.

These tourists, students and citizens, today find themselves crossing a citadel of illegality, areas abandoned to abuse, dismal during the day and to be avoided at night. All this while the increase in young population and tourism, in turn, will be boosted by other planned interventions.

Restoring legality, starting from Monte dei Cocci (and it is for this reason that a participatory process has been initiated), is a responsibility that local government can and must take on. The territory where these incidents occur, and where there is a vast concentration of illegal construction, is the Prati del Popolo Romano, an entirely public property of the Municipality of Rome.

Another piece of equipment that the territory lacks but presents a particularly urgent need for is a system of bike paths serving university access. The Ministry of Transport indicates cycling as the preferred mode of transportation for home-to-school routes. Research and funding have been dedicated to this topic. In particular, a study conducted by the RFI (Italian Rail Network) Stations Directorate and adopted by the Ministry, analysed the relationships and connections between railway stations and universities. The results highlighted three key factors for the development of this connection: the train is the primary mode of collective transportation used by students, bicycles are used for daily commutes, especially by young people between 18 and 24 years old. Furthermore, over 80% of university campuses have at least one railway station within a 3-kilometre radius.

For the Testaccio area, there are two railway stations, Ostiense and Trastevere. They are located along the main axes of the railway system, therefore part of the primary network of national transport, and concerning the former Slaughterhouse, they are well within a distance of three kilometres. Additionally, there is a subway station (Piramide) and the terminus connecting Rome to Ostia and the sea.

The equipment of the territory is particularly favourable to this type of connection, especially considering that Testaccio, due to its cul-de-sac shape, could not support an increase in private motorized mobility. However, the analysis of the table above reveals how far we are from a modern service logic for existing bike paths: diversifying essential home-to-school and home-to-work travel across multiple vectors.

The first observation concerns connections to railway stations. They are entirely to be built, as mentioned earlier.

In their current distribution, the paths are fragmented and disjointed, squeezed into minimal residual spaces, and at the nerve centres of urban traffic, for example, in the Ostiense square, they lead to fraying, causing anxiety and insecurity for cyclists. The paths are mainly used on Sundays, for sports or leisurely walks, and thus outside a service logic aimed at not only reducing traffic congestion but also the burden of work and study days for workers and students (see map 'Location of interventions within the boundaries of the Testaccio district').

A contribution to the reconsideration of the cycle and pedestrian network comes from the Teverever project, whose bike path passes under the rear facade of the former Slaughterhouse. In this segment, we find a particularly interesting point of ascent. These are not the stairs we know but a convenient and wide driveway ramp that also allows service vehicles to descend onto the quay. It's a rare opportunity. Another similar ascent is found after five kilometres, at the De Pinedo stop, behind Piazzale Flaminio. Teverever covers the stretch from Ponte Marconi to Ponte Garibaldi and will be implemented in successive phases with the arrangement of facing riparian parks, planting of suitable trees, and recreational facilities, all aimed at the construction of a new bike path. The particular problem with this intervention, as mentioned earlier, is the ascent to the city level. The stairs that the embankments are equipped with, besides being untouchable because they are protected, were originally designed for mere service functions and not for visitors. Therefore, they cannot offer satisfactory access today, especially for elderly people or individuals carrying loads, such as wheelchairs, bicycles, etc. The issue is further complicated not only due to constraints but also for safety reasons. The banks must be kept clear to allow for the discharge of any floods. Thus, the idea that the Tiber, in its current state, can contribute to traffic alleviation in the city appears entirely abstract. This potential is yet to be built, but it must be admitted that at least Teverever moves in this direction, especially when compared to the GRAB (Grand Bike Ring), whose route passes nearby, along the Circus Maximus. In the segment of interest here, between Ponte Sublicio and Ponte Testaccio, a radical cleaning of the quay is already underway.

These cleaning works form the basis for further considerations. In the approximately one-kilometre stretch between the two bridges, thirty encampments of people in a state of utter abandonment were found among the bushes. This is just the most recent occurrence. Another group in similar conditions was removed from the former Slaughterhouse when the construction site for the Academy of 'Belle Arti' opened. During the reclamation of the former Testaccio camp, another group of homeless individuals was entrusted to social services, while even today, in the abusive settlement of Borghetto Caselli, heterogeneous groups of people live in conditions of fragility or profit from illicit interests (rents, illicit activities) regarding the use of shacks. It is the exceptional nature of the intervention in Testaccio, which assigns new functions to a plurality of areas and buildings in a state of abandonment that highlights the absence of governance of the changes. This barrage of interventions on the homeless, with all the humanitarian implications it entails, is a significant anticipation of the future traumas to which citizens will be subjected.

Alongside all these issues, essentially a list of priorities on past internal matters and adjustments made urgent by upcoming interventions, there is a theme that is advancing independently. It is the dual issue of energy protection and sustainability.

In the city of Rome, the construction of some energy communities has been initiated, an initiative that involves public buildings which, with appropriate equipment, can become producers and sellers of electricity. Even individual condominiums can set up and produce energy. A complex system of opportunities and financial contributions links these two areas of intervention, but the important result is that a domestic reflection is opened, among condo owners, students and citizens, on the value of energy and its connection to the environment. One of these communities is also being built in Testaccio.

On a different front, the theme of sustainability in new constructions is moving forward. Tender procedures, especially for projects funded by European grants, carry a new concept. Companies must take responsibility for the regeneration of the work at the end of its use cycle (the motto is 'from cradle to cradle') and anticipate the recovery methods in their offers. In the case of Testaccio, the multiplicity of interventions provides an opportunity for a coordinated analysis of the different minimum environmental criteria (CAMs). The genuine stakeholders, the citizens, could assess the benefits of their combination in their living and working environments. Thus, an energy and environmental awareness is born, which can shape a new dialogue and a new frontier of participation.

3.6 Towards the New Arrangement

The layers of projects in Testaccio, accumulated over decades, can represent a wealth of wisdom from which to draw indications and suggestions for abandoned areas, and consequently, to engage with the new functions coming in. Among all the contributions, we will select at least two.

The first inspiration comes from Raffaele De Vico's project for Parco Cestio, a journey that began in the second half of the 1920s and was completed in 1939. Documentation from 1926 already indicates that the area hosts a precious visual reference between the Pyramid, Porta S. Paolo, and Monte dei Cocci. The places were intended to become an oasis of peace in the heart of the city; they could have contained a sports field (that of Roma would be operational from 1930), a swimming facility, a space equipped for the De Amicis professional institute, and the construction of a fire brigade barracks. De Vico's park followed this general layout (Fig. 3.2).

If we consider the current situation, the intended use as an urban park that may also include sports facilities is still feasible. This initial design indication, the reversion to a green oasis for a degraded area, in the name of the tradition of play and sports stemming from the historical uses of the Prati del Popolo Romano, is a compelling approach. The only incongruous element is the Caselli settlement, an illegal settlement that started in the post-war period as a response to the housing emergency, a trace of no value incompatible with the context.

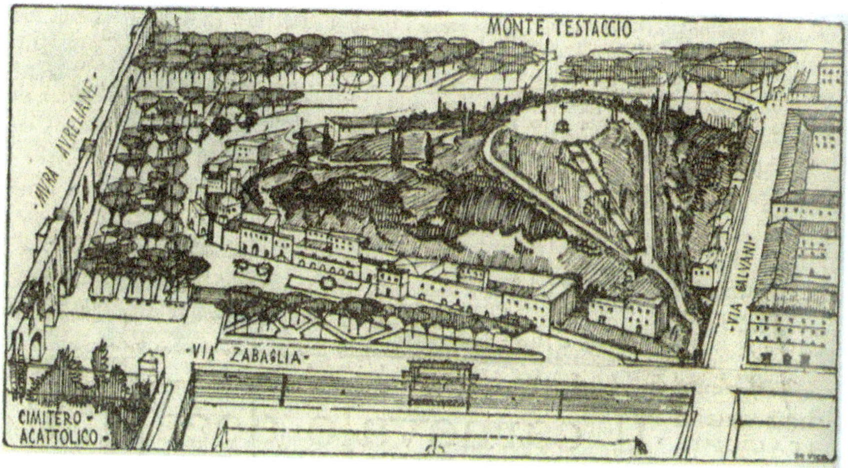

Fig. 3.2 Project by architect De Vico for the new park, from 'Anche il Testaccio' 1931

De Vico's park also underlies the more recent Caruso-Aymonino project, drawn up in the mid-1980s. The design proposal places the new football field in a vast space freed up between via Marmorata and via Zabaglia. Around it there are arranged residences, schools (later built), and workshops. The settlement is eradicated. The sports facility serves as a monumental visual telescope between the Resistance Park and Monte dei Cocci, confirming the characteristics of the aforementioned area.

The new university functions assumed by the former Slaughterhouse today allow for a further elaboration of all these indications. The vocation as a park is confirmed, as well as a new role for sports facilities in homage to the history of AS Roma football club, which may include gyms and even the facilities of the 'Elsa Morante' school complex. The additional design indication may consist of the preparation of a modern Sports Campus serving the student population and all citizens. Through this approach, the construction of the new university campus can drive a design solution for the degraded area indicated, giving new purpose to the unfinished projects of Testaccio. The recovery of buildings around Monte dei Cocci can provide space for housing, artist studios and craft workshops, offering a landing place for any such functions that persist today in the Caselli settlement.

The excavations conducted in the area, the recovery of the Emporium and the associated service areas, the progressive affirmation of the idea of the Aurelian Walls as a linear park, the reinterpretation of Parco Cestio, the enhancement of the historic cemeteries, the university functions—all these elements can lead to an arrangement of great prestige and beauty. The binding element of all these suggestions will be the network of cycle paths, made up of comfortable, safe routes usable without anxiety and worry, open towards the GRAB, the CaRMe facility and the Teverever track, with connections to railway stations and the metro.

Finally, it will be necessary to provide predictions regarding changes to the commercial system and service provision.

3.7 Final Remarks

It is evident from the data presented above the relevance of the intervention involving the entire Ostiense and Valco San Paolo area, even though the centre of gravity is located in the Testaccio district. The exceptional nature of the case is certainly due to the PNRR (a substantial volume of funding and implementation of works by 2026), but it is also based on the 2025 jubilee program and confirmed by the recovery of portions of the former Slaughterhouse of interest to the Municipality of Rome, a process that started in 2003 but whose construction sites now represent the highest financial investment and the capture of the most qualifying functions of the entire area. Therefore, it should not be surprising if the heterogeneity of the programs, the dissonant timing, and the simultaneous arrival of the realization of projects matured in different times and contexts pose a new and unprecedented problem: finding compatibility retroactively with an idea of urban layout that develops and surpasses the indications of the PRG and the projects dependent on it and points in ways to be found in a green, accessible city, gearing up for change.

Consider the criticisms from the outset of the PNRR being not a plan but a list of macroeconomic investments dropped onto the territory (a reserve also placed on Caput Mundi, of which, on the contrary, the 'scrappy' character of the majority of projects has been emphasized). These concerns did not seem unfounded, as the dedicated Commission of the Municipality of Rome drafted a Report identifying projects complementary to those already financed by the PNRR. The Commission examined individual projects (the so-called flagship projects) against which further projects were simultaneously identified to strengthen the planned investment or integrate it with interventions in areas not covered by the PNRR or with complementary works. All this to retrospectively obtain a program of investments more responsive to the needs of citizens and with a better social impact[2].

The above-mentioned CaRMe, is the most convincing proof and example of the reversal of approach: PNRR projects acquire justification not because they are respectful of the original program, but through the amalgamation of existing diverse initiatives and the recognition of future needs. The CaRMe project proposes a new vision of the Central Archaeological Area, projecting it into the future, yet finding within it the interventions (even the tiny ones) owed to Caput Mundi.

This foresight and this sensitivity to the effects of works on the territory cannot be limited only to the PNRR. Returning to the Ostiense area and the Testaccio district, a similar approach should be applied here as well, although the PNRR quota does not play a predominant role. Let's analyse the proposed additional intervention that the aforementioned commission makes for the Testaccio area. It concerns the stretch of the Aurelian Walls from Porta S. Paolo to the Tiber, concerning the redevelopment of the pedestrian area along Campo Boario Avenue (project, already existing, estimated amount of €3.00 MIL). The Linear Park of the Aurelian Walls is the 'flagship project' to which reference is made. But without falling into the trap of comparisons, in the

[2] Capitolina Assembly Special Commission PNRR. Report No. 4 activities as of December 12, 2023

Ostiense area and Testaccio, even the intervention in the former Slaughterhouse could have the characteristics of a 'flagship project'. It would therefore deserve the same approach and should involve the entire set of programs operating in the area, including the PNRR.

Throughout this paper, themes have been anticipated on which concurrent design should focus: a modern network of cycle paths, the redemption from the illegality of the vast areas surrounding the former Slaughterhouse. Further themes are to be addressed. We hope that these criteria are adopted in the revision of the Urban Project Ostiense-Marconi.

3.8 Conclusions

Another conclusion that can be added to the end of this article is that the case study presented confirms the importance of adopting a bottom-up approach in the management and transformation of urban areas, as in the case of the Testaccio district, by integrating self-sufficient and sustainable aspects into the regeneration project. The various interventions described, despite presenting significant challenges related to historical, cultural and urban constraints, demonstrate how it is possible to create an urban environment that promotes energy self-sufficiency, sustainability and active community participation.

In particular, the creation of energy communities and the adoption of sustainable mobility solutions, such as bicycle lanes, not only contribute to reducing environmental impact but also foster a cultural shift regarding energy consumption and the use of urban spaces. These aspects are crucial for building resilient cities, where the community can be directly involved in decision-making processes, improving the quality of life and liveability of the neighbourhood.

The integrated approach that combines the management of historical heritage, such as the enhancement of the Aurelian Walls, with sustainable urban regeneration, represents a development model that, if implemented correctly, could become a replicable example for other Italian and international cities. The connection between universities, public spaces and green infrastructures not only helps reduce congestion and pollution but also creates opportunities for socialization and inclusion.

Sustainability and Participation: A Model for the Future

In addition to these fundamental themes, another key aspect emerges: the dual issue of energy protection and environmental sustainability. The regeneration project in Testaccio is part of a broader context of growing attention to sustainability, with the introduction of energy communities where public buildings and condominiums become producers of electricity. This initiative, which allows energy to be produced and consumed in a shared manner, sparks a deep reflection on the value of energy and its relationship with the environment. In particular, the energy community in Testaccio becomes an example of how, through citizen participation, it is possible to build an energy self-sufficiency system that strengthens the bond between residents and the territory.

In parallel, the sustainable approach extends to new developments, with an increasing commitment to ensuring that interventions respect long-term sustainability principles. Tender procedures, particularly those related to European funding, encourage the adoption of construction practices that not only reduce environmental impact but also foresee the regeneration of the works at the end of their lifecycle. In this context, the plurality of interventions in Testaccio offers the opportunity to apply the minimum environmental criteria (CAM) in a coordinated manner, leading to integrated resource management that benefits both the environment and the quality of urban life.

A New Model of Participation and Inclusion

However, the real challenge of this project is the ability to actively engage the community in all these initiatives. Participation is not limited to simply benefiting from the changes brought about by urban regeneration but becomes a central element in defining energy and environmental policies. In this context, the Testaccio project can serve as a model for a new type of civic participation, where citizens are called not only to benefit from the changes but also to be protagonists in defining and implementing the most suitable solutions for a sustainable future.

The creation of energy and environmental sensitivity among citizens, combined with their active participation in daily decisions regarding energy use, could represent the foundation for a new model of the city. Sustainability practices, indeed, are not limited to managing natural resources but extend to promoting a cultural shift that sees citizenship as the guardian of collective well-being and actively contributing to improving their environment.

In summary, the adoption of a development model that considers environmental sustainability, social regeneration and civic participation, as outlined in the Testaccio case, has the potential to transform the area into a virtuous example of self-sufficiency and sustainability. It is hoped that future projects, such as the Ostiense-Marconi Urban Project, will embrace these principles, shaping an urban environment that harmoniously and durably meets the needs of the community and the territory.

References

Arnstein, S. R. (1969). A ladder of citizen participation. *Journal of the American Institute of Planners, 35*(4), 216–224.

Beatley, T. (2012). "Green urbanism: Learning from European cities." Island Press.

Bellone C. (2021), I Prati del Popolo Romano. La partecipazione come sperimentazione per il progetto dello spazio urbano, Franco Angeli,. ISBN 9788835122555, On line 8835122554.

Bellone, C., & Trevisani, P. (2020). Dal processo partecipativo al Progetto Urbano. Il rione Testaccio (Roma). *Urbanistica Informazioni, 292*, 50–52, ISSN: 0392-5005.

Bellone C., Trevisani P, & Colonna A., (2018). "Il recupero urbano attraverso esperienze di "partecipazione": piazza Testaccio a Roma". *Urbanistica Informazioni*, vol. 278–279, ISNN 0392-5005 (rivista stampa) ISSN: 2239-4222 (rivista online).

Bellone, C., Trevisani, P., & Colonna, A. (2019a). The participation role of citizens in "piazza Testaccio" regeneration project in Rome Historical City center. In: Ruggiero, C., Arslan, H., Gianturco, G. (Eds.), *Contemporary approaches in social science researches* (pp. 283–286), E-BWN, ISBN: 978-83-953142-1-6 con peer review.

Bellone C., Trevisani P., & Colonna A., (2019b). "Il processo partecipativo per l'ex campo di calcio della Roma nel rione Testaccio". *Urbanistica Informazioni, 283*. ISNN 0392-5005 (rivista stampa) ISSN: 2239-4222 (rivista online).

Bellone, C., Trevisani, P., & Colonna, A. (2020). Piazza Testaccio: esperienza di partecipazione nel recupero di uno spazio urbano nel centro storico di Roma. *Archivio Di Studi Urbani E Regionali, 128*, 64–79. ISSN: 1971-8519. online ISSN: 0004-0177.

Harvey, D. (2008). The right to the City. *New Left Review, 53*, 23–40.

Jacobs, J. (1961). *The death and life of great American cities.* Vintage Books.

Documents

Assemblea Capitolina Commissione Speciale PNRR. Relazione n. 4 attività al 12 dicembre 2023

Deliberazione Assemblea Capitolina n. 44/ 32022

Deliberazione Assemblea Capitolina n.84/2022

DPCM 08 giugno 2023

Piano Nazionale di Ripresa e Resilienza PNRR approvato con decisione del Consiglio ECOFIN del 13 luglio 2021. Missione 1 componente c3, investimento 4.3 "Caput Mundi. Next Generation EU per grandi eventi per grandi eventi turistici" DPCM 15 12 22

Capitolina Assembly Special Commission PNRR. Report No. 4 activities as of December 12, 2023

Capitolina Assembly Resolution No. 44/32022

Capitolina Assembly Resolution No. 84/2022

(DPCM) of June 8, 2023

National Recovery and Resilience Plan (PNRR) approved by the ECOFIN Council decision on July 13, 2021. Mission 1, component c3, investment 4.3 "Caput Mundi. Next Generation EU for major tourist events" DPCM of December 15, 2022

Chapter 4
'Cartographies' of the Landscape of the Interwar Linear Park of Fokionos Negri Street in Athens Through Literature and Art

Georgia Eleftheraki, Maria Markou, and Anna Maria Visilia

Abstract In our time, the preservation of urban heritage is 'a condition sine qua non of the cities' development'. It is an imperative action that 'can better be framed within the larger goals of overall sustainable development through a landscape approach' (UNESCO, Recommendation on the Historic Urban Landscape, 2011). Such an approach consists of identifying and managing historic buildings or areas within their broader urban, natural and cultural contexts, to better comprehend the layering of cultural values, social practices and collective representations, through which the city 'tells its own story'. Our ongoing PhD research is about the interwar urban landscape of Fokionos Negri Street, a green space that was created in the interwar years over a stream and along which leisure, commerce, services and housing coexist, forming the heart of Kypseli, one of the oldest and most densely populated neighbourhoods of the capital of Greece. Unlike most neighbourhoods in the historic centre, Fokionos Negri retains several of the high-rise apartment buildings of Athenian modernism and, after being pedestrianized in the 1980s, has been transformed into one of the rare linear parks in Athens, which remains particularly attractive for the surrounding residential area. We are studying its transformations through archival research, field research and observations, personal interviews, various artistic and cultural 'narratives' and Press. This paper focuses on a diachronic psycho-geographical and mental 'mapping' of the linear park through literature and art, a methodology that enables us to understand how society perceives the identity and physiognomy of the area in different periods. The aim is to contribute to the documentation, evaluation and highlighting of the special historical and cultural values carried by the insufficiently protected urban

---

G. Eleftheraki (✉) · M. Markou
School of Architecture, National Technical University of Athens, 42, Patision str., Athens, Greece
e-mail: elefgogo@gmail.com

M. Markou
e-mail: mmarkou@arch.ntua.gr

A. M. Visilia
School of Design, Architecture and Art, North Dakota State University, 1340 Administration Ave., Fargo, ND 58102, USA
e-mail: annamaria.vissilia@ndsu.edu

F. Naselli et al. (eds.), *Self-Sufficient Cities and Sustainable Habitats*, The Urban Book Series, https://doi.org/10.1007/978-3-032-10337-6_4

architecture and natural heritage of the early twentieth century in and around the central areas of Athens.

Keywords Interwar urban heritage · Historic urban landscape · Linear park · Fokionos Negri street · Kypseli · Cartography

4.1 Introduction and Methodology

Urban heritage, in its tangible and intangible form, is for humanity a social, cultural and economic asset, and a key resource in enhancing the liveability, economic development and social cohesion of urban areas. The principle of sustainable development provides for the preservation of existing resources, the active protection of urban heritage and its sustainable management as a condition sine qua non of development. Urban heritage conservation strategies must be integrated and framed within the larger goals of overall sustainable development, through a landscape approach, that considers historic buildings or areas within their broader urban, natural and cultural contexts, as they are shaped by human interaction through time. Such an approach is aimed at preserving the quality of the human (natural and cultural) environment, enhancing the productive and sustainable use of urban spaces, while recognizing their dynamic character. Thus, it provides tools to manage physical and social transformations and to ensure that contemporary interventions are harmoniously integrated with heritage in a historic setting and consider regional contexts. The historic urban landscape approach draws lessons from the traditions and perceptions of local communities, while respecting the values of the national and international communities (UNESCO, 2011).

The design of sustainable cities should be based on the strategy of integrated residential development, promoting coherent urban mixed-use forms. It is, also, vital to secure and protect significant open spaces and urban green spaces and to create urban green corridors. Another important parameter is to strengthen policies for historic centres and central areas of cities, as well as to implement detailed urban design and improve the aesthetics of the city. A sustainable city should also promote economic development and respond to the basic needs of its inhabitants, while creating sustainable living conditions and well-being for all. This holistic approach can contribute to the sustainability of the city and enhance its resilience, if it also includes the dimension of environmental protection, which means strengthening natural systems and managing the impacts of climate change. Urban heritage conservation, which is based on historical values of the city, also plays an important role, as it '*is a central component of a development agenda that promotes inclusive, safe, resilient and sustainable cities*' (Karlstrom, 2014; UNESCO, 2016). A fundamental feature of urban centres is their historicity, therefore the ability to recall memories, something that fascinates people, as cities "narrate" through their buildings and structures. The historic urban landscape [HUL] includes all tangible and intangible elements inherited from history, tradition and human creation (Stefanou, 1988), as well as from present-day reality, and represents an entity that humanity has a duty to preserve without, however, interfering with the irresistible force of change that is the city itself. UNESCO's approach to the management of historic urban landscapes is holistic, integrating the objectives of urban heritage conservation with those of economic and social development (Eleftheraki, 2022).

Our on-going research project[1], argues that such a holistic approach to the historic urban landscape is needed in Athens. The current legal framework protects and promotes the architectural heritage of the city centre for historical and political reasons, valuing it as an archaeological treasure of significant neoclassical interpretation. However, it omits to include the protection and preservation of important buildings and open spaces of the twentieth century, which complete the image and identity of the city as modern monuments, interpreting the mechanisms of its evolution and the ways in which the people appropriate the city in their everyday life (Eleftheraki, 2022; Eleftheraki et al., 2023). Emphasis on the neoclassical past often obscures people's understanding of recent architectural and social history, but also of how contemporary history relates to different places in the city and different groups of citizens. Furthermore, it limits the possibilities of studying and protecting both buildings that can be characterized as monuments of modern architectural heritage, and urban open places or architectural ensembles that are in also linked to collective memory. The result is that many remarkable twentieth century buildings and complexes are altered or even destroyed (Fig. 4.1). Taking as an example the linear park of Fokionos Negri Street, in the district of Kypseli in Athens, our on-going research project is trying to develop a method of studying this historical urban landscape. Our method focuses on the period of 'Athenian Modernism' and utilizes, among other things, artistic approaches to the urban landscape, in which this paper focuses on.

4.1.1 *The Arts and the City*

Descriptions of the city of Athens existed long before scientists, and, in our time, are carried out by countless littérateurs who approached and recorded the city, its name and its inhabitants in the most accurate way. Such place-based narratives of Athens, formed through daily routines, observations, and memory, have been discussed in cultural writing on the city (Vatopoulos, 2002). It has been argued that every great writer today is an author of the city, and, in fact, there is not a single creator from the nineteenth or twentieth century who did not write about the city (Mongin, 2003). Homer's 'Iliad' (Ilion) is also a 'poleography'. Of course, '[writing] *cities are often the product of beautified memories and romantic writing*' (Pantaleon, 2006). However, '*literary cities*' are an inseparable, important and vibrant component of '*urban chronicles*'. Urban narratives and literary descriptions, along with poetry and

[1] The PhD research in progress titled "Diachronic interpretations of interwar urban landscapes. The "modern" apartment buildings of Fokionos Negri".

Fig. 4.1 Views of the two-story interwar building on Ioannou Drosopoulou and Syrou Street (first parallel to Fokionos Negri Street) in chronological order, adapted by Georgia Eleftheraki based on N. Karatziou photo of 2023 (https://www.lifo.gr/stiles/optiki-gonia/katedafiseis-istorikon-ktirion-mia-mikri-niki-se-mia-megali-mahi) and on the Google Maps/3D, Ioannou Drosopoulou and Syrou Street (2009 and 2011). The building has been closed in recent years and it is going to be demolished. For the time being, the demolition has frozen, following the efforts of the Hellenic Society for Environment and Culture (ELLET). The elevated ground floor of the building has, over the years, housed a variety of uses, such as commercial store (2009), restaurant, etc. Since the submission of the manuscript, the building has been demolished (mid-2025), highlighting ongoing pressures on interwar built heritage

drama, have the power to create powerful intellectual performances. Thus, through storytelling a mental map of the city can be created. The artistic representations of the cities represent an alternative cross-cultural approach of the creative industry, thanks to the inventive and innovative approaches of the artists in the fields of painting, photography, cartoons, comic strips and films.

Furthermore, as the city of Athens is constantly mentioned and referred to in a plethora of everyday written sources, newspapers and journalism have the ability and power to contribute to its reputation. Cities exist as spatial units but, at the same time, can be found and accessed from reference sources such as encyclopaedias, geographical dictionaries and research papers or monographies. The cities also appear on the pages of newspapers and magazines or on television images. One could say that the city as a human construct competes the image formed through its references (Rentzos, 2006). After all, the place consists not only of its spatial data, but also of its chronicles and the 'Intellect' and 'Myth' created by its people's activity and work (Stefanou, 2017). Visual art and literature represent and describe in creative and eventful ways cities teeming with people. Architecture and film are closely linked to contemporary culture. As a form of imaginative journey, the moving image shares the dynamics of architectural walking and can create its own movement (Webber & Wilson, 2008). The interrelationship between cinematic art and structured urban space creates an urban odyssey that combines reality and imagination.

4.1.2 The Linear Park of Fokionos Negri Street in Athens

Kypseli is one of the oldest and most densely populated districts of Athens (Vaschenhoven, 2003), which stretches (Fig. 4.2) east of Patision Street and north of Pedio Areos Park (Field of Mars),very close to the neoclassical core of Athens. Fokionos Negri is Kypseli's main street and constitutes, together with Konstantinos Kanaris square, the heart of the district, a street-linear park, where commerce, leisure and services coexist with residency. Created during the interwar period, when very few green spaces were planned, it was from the beginning a highly desirable place to live (Fig. 4.3), attracting upper- and middle-income groups who invested in modern apartment buildings. Thus, it is still hosting many buildings characteristic of 'Athenian Modernism'.

Fokionos Negri Street creates an enlarged urban void in the densely built fabric of the district, with the minimal disturbance from cars, a rare quality for the Athenian centre. The high vegetation makes the sidewalk look like a 'green torrent' crossing the

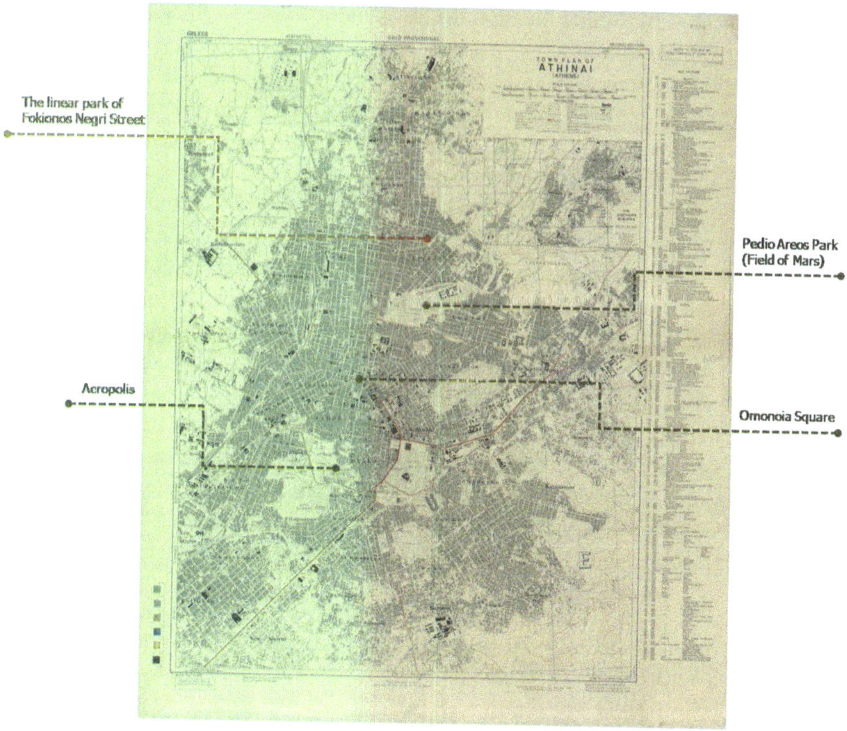

Fig. 4.2 Town Plan of Athens showing the location of Fokionos Negri and other important sites, adapted by Georgia Eleftheraki from Town plan of Athinai (Athens). G.S.G.S. 4457 (London: [s.n.], 1944), Great Britain, War Office, General Staff, Geographical Section; Princeton University Library, Digital Maps and Geospatial Data (princeton-gt54kq43t) (https://geo.nyu.edu/catalog/princetongt54kq43t)

Fig. 4.3 Panoramic view of Fokionos Negri Street and Ano (Upper) Kypseli from the height of Ioannou Drosopoulou Street (1964) © Billy Files (http://vintage-files.blogspot.com/)

neighbourhood (Fig. 4.4). Those characteristics make it highly attractive for living even today (Fig. 4.5). In this location, as throughout Kypseli's district, the typology of apartment building was chosen as the main residential archetype. During the interwar period, buildings of this category followed the design vocabulary of the modern movement and were intended to house wealthy bourgeois families offering a particularly elegant and comfortable style and way of living. In our time, the built environment in Fokionos Negri consists of interwar and post-war apartment buildings (Eleftheraki & Lekai, 2020), with a landmark being the interwar Municipal Market of Kypseli. The preservation against the demolition and the management of the Market has recently created intense controversy about the importance of public spaces and architectural heritage in Athens. Around the 1990s, after a period of disinvestment as the negative impact of peri-urbanization, Kypseli's district attracted immigrant groups and developed a multi-cultural character. Today, it seems to be experiencing its 'redemption story'. The gentrification phenomena that are developing in Kypseli are guided by the myth of Fokionos Negri Street as a 'timeless symbol of Athenian life'. This myth finds its material expression in the modernist heritage of the area, which from the beginning expressed the city's ability and need to assimilate international trends of development. The main objective of this study is to promote the preservation and sustainable management of Fokionos Negri's architectural and cultural heritage.

4.1.3 Methodological Approaches

The research aims to study the historical urban landscape along the linear open space of Fokionos Negri Street, which over time influenced the social and functional evolution of Kypseli's district, if not the Athenian centre. The interwar urban landscape is approached from an architectural, spatial and social point of view, based on extensive

Fig. 4.4 The linear park of Fokionos Negri Street as a "*green torrent*" is crossing the neighbourhood (https://www.millenniumrealestate.gr/en/propertyDetails/1305675)

Fig. 4.5 Modern view of Fokionos Negri Street from the height of Ioannou Drosopoulou Street: (https://bonflaneur.com/athens_points/fokionos-negri/)

archival research, field surveys and observations, personal interviews, and various artistic and cultural 'narratives'.

The methodology of the Ph.D. research includes:

a. Recording and documentation of the interwar buildings of Fokionos Negri, mainly through on-site and primary research, but also from secondary sources.
b. Investigation of the social history and uses of the buildings and the wider area over time, through various and different sources and personal interviews with their residents or patrons of the linear park.
c. Presentation and analysis of the remarkable morphological and typological characteristics of these buildings, as well as the history of their people, generating a folio—'The Small Notebook of the Modern Interwar Apartment Building'—for each of the interwar apartment buildings of the street.
d. Documentation and analysis of the landscape design of Fokionos Negri from the interwar period with historical references to landscape architecture.
e. Creation of a different 'mapping' of Fokionos Negri from the interwar period and the first post-war decades through its projection from various forms of art and culture.

f. Articulation of observations and conclusions on the architecture and landscape heritage of the street, as well as on the social organization, uses and lifestyle through empirical research with interviews.
g. Investigation of the 'integration' of the interwar heritage in the modern urban landscape, examination of the way of its assimilation and influence, through photographic documentation, observation and interviews.

This study focuses on a diachronic psycho-geographic and mental 'mapping' of the linear park through literature and art, which enables us to understand how society perceives the character and physiognomy of the area in different periods of time. The aim is to contribute to the recording and evaluation of the incompletely protected newer urban buildings and natural heritage of the Athens' neighbourhoods, highlighting their special historical and cultural value.

4.2 'Cartographies' of the Landscape of the Interwar Linear Park of Fokionos Negri Street

4.2.1 'The Most Beautiful Street of Athens'

Michaelides[2] *states, 'We have been living Fokionos Negri since the years when it was a stream. Stream, which descended the waters from Turkovounia when it rained, passed under Patision [Street] to Agios Meletios and continued to Kifissos* [River]. *Just above Kypseli's Square were the butts of the Hellenic Army Academy. Then, Kostas Kotzias*[3] *came as mayor. He closed the stream, planted trees, made a lake, built the Municipal Market. The place was beautiful. And we started to melt the soles of our shoes on the walks. When we had little money in our pockets, we sat in the cafe of Petros Lameras. The most common, however, was to be standing next to the kiosk of Kaliakoudas. There were also the big celebrations when there was money in our pockets; we were sitting in the bar of Dritsas for beer. This was rare, very rare …'*.[4]

Fokionos Negri Street, named after the well-known academic and politician Fokion Negris, was established in the 1930s during the mayorship of Kostas Kotzias (1934–1936), following the arrangement of the Levidis Stream. The stream started from the slopes of Aghesmos (Tourkovounia), flowed along Krissis Street in Ano (Upper) Kypseli, crossed Kypseli Square and then the current pedestrian street of Fokionos Negri. The two sides of the stream were joined at two points, with bridges

[2] Panagiotis M. Michaelidis (Athens, 1921). His parents, both Pontians by origin, were refugees from Crimea, Russia. He studied at the School of Physics and Mathematics of the University of Athens. He took part in the National Resistance (1940–44).

[3] Konstantinos (Kostas) Kotzias (Athens, 1892–1951) was a Greek lawyer, journalist, politician, fencer, publisher and writer. He became minister (1936–1941) and mayor of Athens (1934–1936).

[4] Michaelides, P.M. 1991. *Agathoupoleos 7 Small stories from the great Occupation*, Bookstore of "Estia", Athens, p. 28.

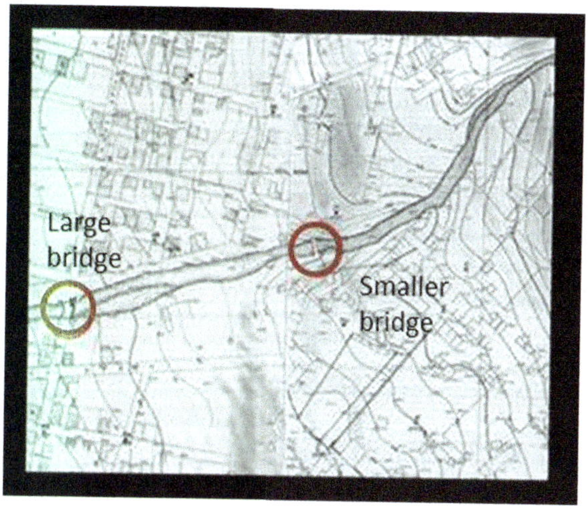

Fig. 4.6 Topographic map of the Levidis stream adapted by Georgia Eletheraki based on D. Lampadarios topographic map of 1927 (Theodosopoulos, 2016)

(Fig. 4.6), a large one at the height of Eptanisou Street (red circle on the left) and a smaller one at the height of Haironeia Street (red circle on the right). In the 1920s, the Levidis Stream, like many other streams in Athens, gradually turned into a source of contamination, dangerous for public health, with stagnant water and waste thrown by residents due to the lack of an organized system for their collection. The increasing urban pressure around the Levidis Stream during the 1930s, combined with the indifference of the authorities, led to its significant degradation, so that the Press of the time characterized its section below Patision Street, outside the boundaries of Kypseli, as '*the awful stream*' (Theodosopoulos, 2016) and pressed for its undergrounding. In the early 1930s, the section of the Levidis Stream in the boundaries of Kypseli, it was arranged and in 1937, the architect Vasilis Tsagris, Director of the Technical Service of the Municipality of Athens, above the pipeline, founded a linear park populated with trees, shrubs, fountains and playgrounds, between two vehicular streets (Fig. 4.7) and the residents started to walk along it (Fig. 4.8).

In 1935, the Municipal Market of Kypseli was built (architect Alexandros Metaxas), which has recently been classified as a monument being an '*excellent example of a public building built during the interwar period*' (Government Gazette 1580/B/16-11-2005). It was erected by the Municipality of Athens, in the aim of relieving the Central Market of Athens on Athena Street (Daniel, 2017). Inaugurated in 1937, the Municipal Market housed mainly food shops and became a reference point for the around neighbourhoods. Thus, it is considered '*a bearer of the history of a type of trade that is nowadays disappearing, as it has been discredited by the generalized commercialization*' (Government Gazette 1580/B/16-11-2005).

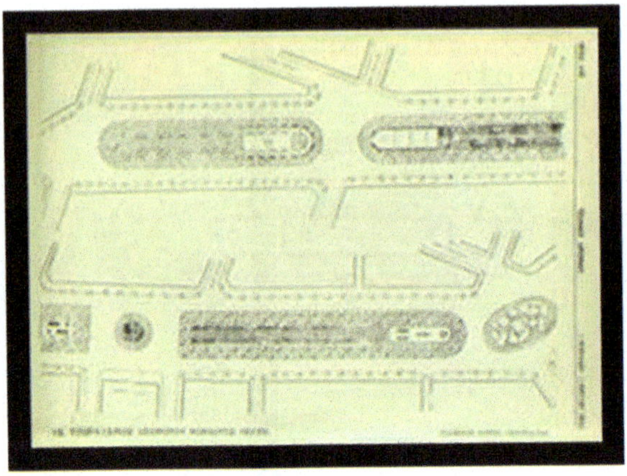

Fig. 4.7 Architectural Drawing Board 1: Embellishment of Fokionos Negri Avenue (Technical Chronicles, 1935, p. 398)

Fig. 4.8 Fokionos Negri Street in the mid-1930s with the Levidis stream boxed in, but not yet covered (© Titika Bournaki Archive, https://www.facebook.com/groups/kypseliotis/, https://www.facebook.com/Kypseli)

In the vignette *Dromopathia* (*Dromopathy*—27 April 1941), the hero of Kostas Varnalis[5] (it is the author himself) chooses to take his walk in Fokionos Negri (instead of the area of Kolonaki where he lives) because it is '*the most beautiful street of Athens*'. This choice takes on greater significance as the vignette was written the previous day—while the Germans were advancing on Athens, on 27 April 1941, knowing that the city (and the entire country) would soon be under occupation for an unknown period.

> A street… heartwarming in every season and… sexy is Fokionos Negri Street in Kypseli. The most beautiful street in Athens. I'm going that way.
>
> - So far away?
>
> - There I will romance for a while, to forget who I am and who you are. It is a unique street in Athens. It has no match. Because it has abundant running water. It is a series of fountains, staggered channels—all with the purest, most abundant and fresh water. In the middle of the road, which is quite wide, a series of small gardens with all kinds of known and unknown

[5] Kostas Varnalis (1884–1974) was a Greek littérateur. He is known primarily for his poems, but also wrote narrative works, criticism, and translations. He was honoured in 1959 with the Lenin Peace Prize.

> plants and flowers. And these little gardens are closed right and left by rows of plane trees and poplars and acacias, as well as by the stepped channels of water, which flows so quietly that you almost think it is still. In one of the pools, where three lion heads continuously pour water, the world's whitest ducks are swimming! And in the middle of the pool is their little house, a kind of miniature pagoda, with two ladders, for these living creatures to go up and down from the liquid element to the solid.
>
> Where Eptanisou Street cuts the road perpendicularly, there is a wide square opening filled with sand. There the children play, roll over, shout, scream, carry with the passion of their age and in all seriousness enjoy the "free consumption of their surplus energy"—as some biologists define play and some philosophers define art.
>
> One thinks that it feels like being in some spa town or in some European city, where the waters are incalculable and the vegetation is luxuriant... So, there I go... a trip! I really need a change of air. Because, as proud as I am in the ears,[6] I constantly hear sirens inside me! Today I will do everything I can, not to listen to either my own sirens or the... others.[7]

4.2.2 The 'Glamour' of Interwar Modernism

Since the 1930s, the '*green avenue*' has acquired apartment buildings, which follow the dictates of the modern movement. However, the area continued at this time to be characterized by a low building density and to have many open spaces. The modern architectural heritage from the interwar era appears in the films of the 1950s, highlighting the street's interwar identity. The emblematic 'Lanaras apartment building' (1938), residence of the well-known family of industrialists from Naoussa, whose name is associated with the beginnings of the Greek textile industry, is presented in Alekos Sakellarios' film, entitled *Neither cat. Nor damage* (*No harm. No foul*) (1955) as the neighbouring residence of the wealthy main protagonists (actors: Vasilis Logothetidis and Ilya Livykou). The secondary protagonist (actor: Lambros Konstantaras) is observed to be waiting for the beautiful Ilya Livykou, in the well-known pastry shop of the time, *Select*, opposite the 'Lanaras apartment building' in Fig. 4.9 (left). The famous apartment building also appears in the movie *The drunkard* (1950) by Giorgos Tzavelas (Fig. 4.10 on the right of the photo), together with its neighboring, very interesting modern 'Tompros apartment building' (1935) on 21, Fokionos Negri Street, on the left of the photo 4.10.

In the film *Lost Angels* by Nikos Tsiforos (1948—Fig. 4.11), the viewer sees part of the facade of the interwar 'Zounis apartment building' (1939), where the central entrance can be observed, which is intended for the bourgeois owners of the apartments. Next to the central entrance, the service entrance to accommodate the working staff in the building, smaller in size, is located on its right side. A large opening on the left side of the main entrance is noticed so that the concierge can keep an eye and control who enters the building. Finally, in a scene from the movie *The Beauty of Athens* from 1954 (Fig. 4.12), in the background on the left we see

[6] This idiom means "hard of hearing". Kostas Varnalis was indeed hard of hearing.

[7] Excerpt from the vignette by Kostas Varnalis entitled "Dromopathia", Proia Newspaper, April 27, 1941. The "other sirens" in the text were those of the war alarm.

Fig. 4.9 Scenes from the film *Neither cat. Nor damage* (*No harm. No foul*) (1955) (provided by Georgia Eleftheraki)

Fig. 4.10 Scene from the film *The Drunkard* (1950). Source: Facebook group "Κυψέλη και Κινηματογράφος" [8]

Fig. 4.11 Scene from the film *Lost Angels* (1948) (provided by Georgia Eleftheraki)

the first, perhaps, interwar modern apartment building on 9, Fokionos Negri and Agathoupoleos Street, the 'Protopappas apartment building' (1933).

[8] We found very interesting film material related to Kypseli on the above Facebook page (Films that have been shot in and around Kypseli or have footage about Kypseli). The creation of this FB page has, as its administrator states, the goal of making the "old Kypseli" known through Greek Cinema.

Fig. 4.12 General view of the junction of Fokionos Negri Street and Ioannou Drosopoulou Street (Kypseli, Athens), showing the linear park, pedestrians, and surrounding buildings, from the film The Beauty of Athens (1954). Source: Facebook group "Κυψέλη και Κινηματογράφος"

During the 1950s and 1960s, Fokionos Negri succumbed to a building frenzy, as the remaining single-family and semi-detached houses gave way to high-rise apartment buildings, on both sides of the street. It is important to note, however, that all these interwar apartment buildings have survived to this day, becoming landmarks of the area, with the most recognizable being the 'Lanaras apartment building'.

4.2.3 The Athenian 'via Veneto'

Fokionos Negri was cosmopolitan, exciting and the liveliest area of the city, day and night. It is no coincidence that in the 1950s and 1960s the Athenians called it the 'Via Veneto of Athens', as a direct reference to the legendary street of the Italian capital. With the same description 'Via *Veneto native in a newer version*', it is sung by the singer Giannis Vogiatzis in the famous song *Fokionos Negri* (1965). Also, it was one of the most expensive neighbourhoods in the city. The modern apartment buildings, the parks, the fountains, the famous bars and above all the 'polite society' that frequented the area, from artists to businessmen and from the rebellious youth of the 1960s to the elegant ladies of the Athenian high society, gave an incomparable glamour to the street. As the architect and NTUA Professor Dionysis Zivas (1928–2018) said in an interview (2006, https://www.monumenta.org/magazine), '*Fokionos Negri is a pre-war achievement*'. In the film *The Beauty of Athens* (1954) by Nikos Tsiforos, the protagonists (actors: Mimis Fotopoulos and Gelly Mavropoulou) meet in Fokionos Negri of the early 1950s, a green street—linear park (Fig. 4.12 and 4.13), which also features original artworks, such as the *Hound Dog* sculpture (Fig. 4.13) by the sculptor Euripides Vavouris, which was installed in 1940. The statue is a full-length marble sculpture, 1.65 in length (Antonopoulou, 2003). The statue can also be noticed in the 1963 musical film *Some people prefer it cold*, written and directed

Fig. 4.13 Black-and-white film still from *The Beauty of Athens*, (1954) showing a couple walking along the tree-lined promenade of Fokionos NegrScene. Source: Facebook group "Κυψέλη και Κινηματογράφος"

Fig. 4.14 Scene from the film *Some People Prefer It Cold* (1963). Source: Facebook group "Κυψέλη και Κινηματογράφος"

by Giannis Dalianidis, but it is placed on a different location (Fig. 4.14). Over time, the sculpture has been placed in various positions, following the remodelling of the park.

4.2.4 A 'Secular' Street

Tasos Athanasiadis[9] thoughtfully remarks, "*The people of the weekend flooded the sidewalks in Fokionos Negri. The cars were going up and down from both sides of the road with an incessant noise. Along with all the smell from the exhaust gases, the smell of the rinsed lemon trees reached their noses. In the Oriental Cafe they found*

[9] Tasos (Anastassios) Athanasiadis (Salihli, Asia Minor, 1913—Athens, 2006) was an Asia Minor writer (mainly novelist and essayist) and academic.

Fig. 4.15 Cinema *Fantasia* (1950) (© *Photo Elite*, https://www.facebook.com/Kypseli)

the frame on the right with the crystal partition half empty. They isolated themselves on one side and ordered toast and orange juice".[10]

Theatres, cinemas and clubs framed the entertainment physiognomy of the area. In 1933, we read in the *Proia* newspaper (22-6-1933) that the theatre of '*our well-known musical theatre artist Mrs Zozo Dalmas*' was inaugurated in Kypseli Square—which defines the 'natural beginning' of Fokionos Negri (https://www.facebook.com/pg/Kypseli/about/?ref=page_internal). At 26, Fokionos Negri Street and Eptanisou Street, the *Media Luz* cinema, immediately after the war (1946–1947), screened '*the finest works of World Film Production*', as advertised in its program. It was later named *Fantasia,* as shown in the *Photo Elite* photography from 1950 (Fig. 4.15). The presence of many bars, theatres and cinemas, as well as the proximity of Kypseli's district to the centre of Athens resulted in the gathering of representatives of the intellectual, artistic and political world in the area. The pastry shops and night clubs of Fokionos Negri were a point of reference for the Greek 'star system' during the golden age of Greek cinema in the 1950s and throughout the 1970s. For example, the famous pastry shop *Select*, which we saw above, started its operation in the late 1940s at the corner of Thiras and Eptanisou Street, opposite the 'Lanaras apartment building' (23, Fokionos Negri and Eptanisou Street). In November 1959, the '*new Parisian*' *Select* (Fig. 4.16) was inaugurated, on the site of the *Media Luz* cinema/dancing restaurant, according to a publication of the newspaper *Ethnikos Kyryx* (17-11-1959), with the presence of '*all the secular Athens, businessmen and politicians*', as it is characteristically written (https://www.facebook.com/Kypseli). It was the hangout spot for a large cinematographic and theatre group (screenwriters, directors and actors), with the most prominent being Nikos Tsiforos (1909–1970), Alekos Sakellarios (1913–1991), Kostas Pretenteris (1926–1978), Christos Giannakopoulos (1909–1963), Orestis Laskos (1907–1992), Vasilis Avlonitis (1904–1970), Nikos Stavridis (1910–1987) etc. (Bouras, 2003). In the 1960s there was also the small *Select*, diagonally opposite, on 21, Fokionos Negri and Thiras, in the 'Tompros apartment building'.

[10] Athanasiadis Tasos.1984. *The Last Grandsons* vol. 1, "Estia" Bookstore, Athens, pp. 179–180.

Fig. 4.16 View of F. Negri from the height of *Select* (1964) (https://www.facebook.com/Kypseli)

The well-known nightclub of the time, *Quinta* (Fig. 4.17), run by the swimmer and polo player of the Olympiakos team, Babis Moutsatsos (1933–2017), opened in 1959. According to a publication of the time, it was the '*most original night club in Athens*' (https://www.womantoc.gr/stories/) and the first music club that is still compared to contemporary night clubs in Athens. It operated at 82, Fokionos Negri Street and immediately became the most famous hangout of its time, which, according to the above publication, '*stays open until morning, thus satisfying the night owls, but also the artists*' (https://www.womantoc.gr/stories/). Well-known names of the international music scene, such as Lucio Dalla (1943–2012), appeared on the stage of *Quinta.*

4.2.5 A Neighbourhood with 'Juvenile Temperament'

The carefree and sometimes explosive adolescence of Kypseli's district in the 1960s is depicted in a multitude of films with a youthful theme and a youthful mood, always against the backdrop of the urban landscape and entertainment venues of Kypseli, the hangouts identified in the collective imagination with 'new' behaviours of the Greek youth of the 1960s. In a scene from the musical film *Some people prefer it cold* (1963), the well-known actors Kostas Voutsas (Kleopas) and Vangelis Voulgaridis (Giorgos) are in one of the youth haunts of the time, *Luna Park* on Fokionos Negri Street, and they play 'football' (Fig. 4.18). We can also see *Luna Park* with its pinball machines in scenes from the dramatic film *O Katiforos* (*The Downhill* ; English release title: *Spoiled Rotten*) (1961) by Giannis Dalianidis (Fig. 4.19).

The 1965 Greek film *The teddy boy of Fokionos Negri* (Fig. 4.20) is indicative of the social identity of the area in the mind of the Greek people, as it is clearly narrated in the movie by the characters representing its haunts and patrons. The so-called 'teddy boys', young people with a provocative appearance and delinquent behaviour, frequented many of its football machines and billiards. In 1958, the Greek Parliament passed Law 4000 as a response to the then growing manifestations of youth denial and reaction. The law remained in force until 1983. The wider importance given by society to this measure can also be seen in the film by Giannis Dalianidis with the title

Fig. 4.17 The entrance of the *Quinta* night club (© Liza Koutsaplis Archive, https://www.facebook.com/Kypseli[We found very interesting and rare photographic material on the above Facebook page. The administrator of this FB page publishes valuable information from the history of the neighbourhood with the aim, as he writes, of introducing "Kypseli of the past" to the public])

Fig. 4.18 Scene from the film *Some people prefer it cold* (1963) (https://www.facebook.com/Kypseli)

Fig. 4.19 Scene from the film *O Katiforos* (*The Downhill*; English release title: *Spoiled Rotten*) (1961) (https://facebook/Kypseli)

Fig. 4.20 The title of the film *The teddy boy of Fokionos Negri* (1965) (https://vimeo.com/)

Law 4000, filmed in 1962 (Fig. 4.21). The scenes of the humiliation of delinquent youths included in this film were filmed on Fokionos Negri Street.

4.2.6 The 'Personification' of the Transformation of the Urban Landscape of Athens

Since the mid-1980s, there has been a movement of mainly new households to the north-eastern and south-eastern suburbs of Athens. However, the older residents remain in the area, especially on the upper floors and larger apartments of the apartment buildings. Due to contemporary Greek cinema moving into another era, Kypseli's district gradually lost its potentialities and glamour (Eleftheraki, 2021).

Fig. 4.21 Scene of the humiliation of delinquent youths in Fokionos Negri from the film *Law 4000* (1962) Source: Facebook group "Κυψέλη και Κινηματογράφος"

Vaiou and Lafazani remark, '*The 1990s is a turning point for Kypseli, as the mass settlement of immigrants reverses the declining trends of the population, the average household size and the age pyramid. The "flight" of the old residents is beginning to be compensated by the establishment of large and young immigrant households*' (Vaiou & Lafazani, 2015). The author Menis Koumantareas (1931–2014), who was also a resident of Fokionos Negri, noted at the beginning of the twenty-first century, "*Maybe in the past, gurgling waters flowed on this street, maybe it was a meeting point for some artists, maybe the bourgeois were at their peak then. But even now, how interesting is the pedestrian street and how much variety has*'. (https://www.mixanitouxronou.gr/i-fokionos-negri-otan-itan-rema-poios-itan-o-filergatikos-ypourgos-fokion-negris-pou-edose-to-onoma-tou-stin-plateia-kai-anakalypse-mia-vythismeni-poli/).

In 2015 the magazine *Flâneur,* which is published in Berlin with texts in English, chose to pay tribute to Fokionos Negri Street (Fig. 4.22) because it embodies the transformation of the urban landscape of Athens in the recent decades. In the same year, a new resident of Fokionos Negri explains his decision to live in the area. '*I chose to stay here* [Fokionos Negri Street] *because it is the most interesting street in Athens, because it is in the city center, but it is no longer considered part of it. Every day you walk along it you discover new things, well hidden, you must look to find them*' (Kyriazis, 2015; https://www.lifo.gr/print/geitonia/o-antonakis-zei-sti-fokionos-negri-ton-oraiotero-dromo-tis-athinas). In addition, the online magazine *Pronews* in an article from February 2018 argues, '*The Street personifies the transition from the "Athenian urban scene" of the past decades to a colourful and cross-cultural mosaic*' (http://www.pronews.gr/istoria/358060_fokionos-negri-deite-ti-metallaxi-toy-astikoy-topioy-stin-athina-foto). Indicative of the importance of urban green spaces in the everyday life of current residents is the opinion expressed about Fokionos Negri Street by photographer G. Malekakis (2018), a long-time resident who returned after several years. Malekakis attests, '*Fokionos Negri plays a dominant role, a tree-lined pedestrian street unique to Athens, a "rambla" that connects Kypseli Square with Patision. In this way, it creates the urban planning*

Fig. 4.22 Issue of the magazine *Flâneur* dedicated to Fokionos Negri (http://www.pronews.gr/istoria/358060_fokionos-negri-deite-ti-metallaxi-toy-astikoy-topioy-stin-athina-foto)

basis for a high quality of everyday life. Beyond that, however, it effortlessly gives off glamour from its urban past' (https://www.womantoc.gr/men/article).

Today, Kypseli's district is an amalgam of old residents, immigrant families, Erasmus students and Airbnb tourists (Fig. 4.23, 4.24 and 4.25). However, it is also beginning to attract new residents, artists, young adults and couples, as well as many lawyers due to the proximity to the Juridical Courts of Evelpidon, etc. These new social categories are returning from the suburbs to their parents' abandoned houses, especially if they are on the coveted Fokionos Negri, to discover a so lively neighbourhood they could have never imagined. Also, another category of new residents are moving to Kypseli from other districts of Metropolitan Athens, due to the dynamics that the area acquired during the Covid-19 pandemic period, with the well-known 'corona parties' that brought it, again, at the forefront of unique youth entertainment. As Marinakis observes, the district at its current form '*constitutes the visual canvas on which the majestic apartment buildings of old Athens embrace the classic haunts, the modern bars and the art spaces that emerge full of vision and courage to deliver us a neighbourhood of contrasts that composes and summarizes the liveliest contradiction of the city*' (Marinakis, 2023). Finally, Tsitsopoulos insightfully summarizes the current feelings and social vibes spotted on Fokionos Negri, when he argues, "*Kypseli, as a living museum of itself, a supreme aesthetic place, a carrier of history, a window to tomorrow. Love, birth, life and in between till the death journeys and music, kisses, dances, friendships, sprees, readings and bodies in ecstasy to feed the wheel, to multiply hope. The hope of the new and the expectation of beauty are what will save the world emerging from the life-giving ashes of the old ideals*" (Stefanos, 2021).

4.3 Discussion

From a methodological point of view, in the occasional artistic approaches to the Fokionos Negri area, we can find valuable testimonies and assessments of the way people appropriate the space, both practically and symbolically. Also, it is worth

Fig. 4.23 Contemporary image from the linear park of Fokionos Negri Street (cafe seating along the promenade). (© Eleni Fotiou,Ζήσε τη γειτονιά μαζί με τα παιδιά... στην Κυψέλη)

Fig. 4.24 Contemporary image from the linear park of Fokionos Negri Street (children's play area within the park). (© Georgia Eleftheraki) Ζήσε τη γειτονιά μαζί με τα παιδιά... στην Κυψέλη)

noted the importance attributed to the area in relation to the wider mutations of the Athenian space. Specifically, cinema and architecture maintain a two-way relationship, as the first one expresses the imaginary version of the real, while the second one expresses the real version of the imaginary. Through this relationship between the two arts, the possibility of creating a wandering in the structured space is formed,

Fig. 4.25 Contemporary image from the linear park of Fokionos Negri Street (fountain surrounded by dense vegetation). (© Georgia Eleftheraki), (Ζήσε τη γειτονιά μαζί με τα παιδιά… στην Κυψέλη)

interweaving reality with imagination. The methodological decision to link the study of the buildings with the stories of their inhabitants on the one hand, and with the stories of the current residents and patrons of the area, on the other hand, seems to provide information of deep empirical and interpretative value. The interwar apartments exhibit architectural qualities, both functional and aesthetic elements, characteristic of Athenian modernism during the interwar period. These elements remain today necessary features included in high-quality residential projects, an important finding that further supports the need for the characterization and conservation of such buildings. By studying interwar architecture, one can get in touch with the economic and social processes and the people that shaped the development of the area of our interest, if not the history of Greece itself. The Interwar has been the model for the commercial apartment buildings of the post-war period. A clear social allocation is identified since the upper floors of were intended for the upper middle class, while the lower floors were intended for lower income groups. Therefore, along Fokionos Negri Street, one can follow the sociological, typological, aesthetic and constructional evolution of the Athenian apartment building from the 1930s to the 1970s, when the post-war housing system began to change. At the same time, one can follow the evolution of the type of functional and social mix that the 'Athenian apartment building' produced. These remarks allow questions related to how the current residents, workers and patrons of the area experience and assess the changes taking place in the urban landscape, as well as to how they perceive the production processes, people and social relations in the city of Athens.

4.4 Conclusions

In the artistic and cultural approaches of Fokionos Negri, fixed points of reference are recognized, such as the Municipal Market of Kypseli, the café *Select*, the linear park and its equipment, etc., which mark a diachronic appropriation of the space by different groups of people through daily use, but also by those who come from further afield. It is also noted that the open-air green spaces, the 'urban oases', which are accessible to the citizen daily and at a short distance from his residence, constitute an important need and, at the same time, a recreation in the densely built-up areas of the centre of Athens, especially in the current context. In addition, the artistic and literary approaches confirm and highlight the importance of urban linear parks for the well-being of citizens. These spaces, thanks to their great length, increase the reach of their immediate environment, allow the development of many activities and enable a greater number of different groups of people to appropriate parts of them and make them their hangouts. The cinematic scenes of the first post-war decades present emblematic activities of Fokionos Negri's linear park, such as pastry shops, strolls, cinemas, night entertainment. At the same time, a new anthropological group is presented, the 'teddy boys', which acquires its own hangouts and marks the area with its own behaviours, coexisting, however, with the high society and the artistic community. The physiognomy of Kypseli served as a source of inspiration for Greek popular cinema of the 1950s and the 1960s, which were the preeminent form of popular entertainment at that time, while simultaneously influencing its overall image. The insistence of film production companies of the post-war period to do 'exterior filming' in Fokionos Negri is precisely due to its recognition as a place representative of the 'Athenian way of life and leisure'. Its appearance in the Greek movies of this period became part of the 'myth' through which the residents of Athens appropriated Fokionos Negri over time.

It is proposed to create an exhibition on the physiognomy of the district of the first post-war decades through images of Greek cinema, literature and Press. The proposed exhibition will, on the one hand, showcase the social history of the neighbourhood and its special historical, architectural, landscape and cultural values, and on the other hand, by assuming the role of 'interpreter' of the urban landscape, it will put into public debate, in various ways and means, its different readings. This exhibition could also be the impetus for the creation of similar exhibitions in other central neighbourhoods of the Municipality or in historic centres of other cities, following relevant research on these areas through art and culture. Such visual art events can contribute to the preservation of urban heritage, as they enable the citizens to understand that '*beyond a monument-based, fully protected and government-funded approach, urban heritage is not a liability, but an asset for cities and communities, contributing significantly to their sustainability*' (UNESCO, 2016).

Our on-going research project fills with a gap in the documentation as the files of the building permits of the interwar buildings of the centre of Athens have been

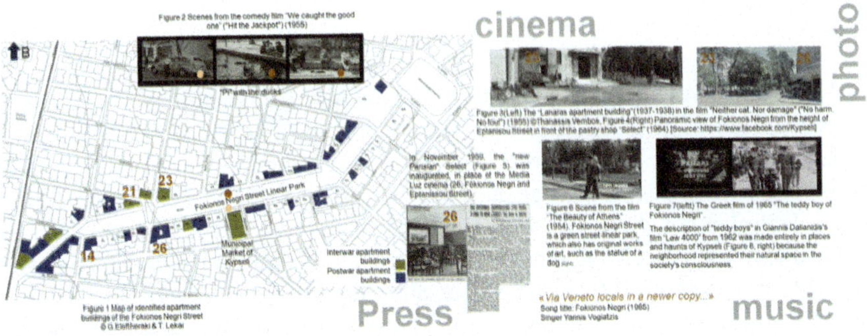

Fig. 4.26 'Cartographies' of the landscape of the interwar linear park of Fokionos Negri Street through art and Press (provided by Georgia Eleftheraki)

destroyed. It, also, contributes to the study of an Athenian central area that was essential for the development of the metropolis and is related to a period in Athens' architectural history that defined the image, functions and society that have long shaped the city centre. The project attempts to record, evaluate and interpret the urban built and natural heritage of the early twentieth century, which has not received, to date, the appropriate attention from the Greek legal framework in order to protect the architectural wealth of this period. At the same time, the project contributes to documenting, evaluating and highlighting the recent architectural and natural heritage of the Athenian neighbourhood and promoting its unique historical and cultural values that should be taken into account when designing the Athenian urban landscape.

By documenting, interpreting and evaluating the architectural, natural and cultural heritage of the early twentieth century, as well as more recent elements of the urban landscape that constitute a perceptible and functional whole of the central area of Athens, we aim to develop a methodology for the study and assessment of the historic urban landscape of Greek cities, which can be used to implement more sensitive and comprehensive conservation and management policies (Fig. 4.26).

Funding Information Funded by the Basic Research Program PEVE 2021 of the National Technical University of Athens.

References

Antonopoulou, Z. (2003). *The Sculptures of Athens. Outdoor Sculpture* 1834–2004.

Athens Magazine. Available at: http://www.athensmagazine.gr/article/retromania

Bouras, D. (2003, Feb 23). The Fokionos of the Greek "star-system". In Newspaper Kathimerini, Seven Days, Kypseli. The urban yesterday, the colorful today, p. 32.

Daniel, M. (2017). The municipal markets of Athens in the interwar period. *Greek Architects*. https://www.greekarchitects.gr

Eleftheraki, G., & Lekai, T. (2020, Jan 24). Urban Modernism in Fokionos Negri in the electronic magazine "Attikipedia_SADAS-PEA". Available at: https://attikipedia.sadas-pea.gr/astikosmodernismos-sti-fokionos-ne/

Eleftheraki, G. (2021). "Urbanography" of a region of Athens (Greece) through forms of art and culture. In *The city and complexity—Life, design and commerce in the built environment. AMPS proceedings series* (Vol. 19, No. 1, pp. 384).
Eleftheraki, G. (2022). The importance of highlighting the multiplicity and diversity of the historic urban landscape. The case of the Fokionos Negri interwar linear open space in Athens. In *13th Italian national institute of urban planning (INU) international study day proceedings on urbanistica informazioni special issue* (pp. 560–562).
Eleftheraki, G., Haniotou, H., & Markou, M. (2023). Highlighting of the Interwar identity of the Fokionos Negri street-linear park: The iconic Lanaras apartment building. *Technical Annals, 1*(3). https://doi.org/10.12681/ta.348079
Karlstrom, A. (2014). Urban heritage. In C. Smith (Ed.), *Encyclopedia of global archaeology*. Springer. https://doi.org/10.1007/978-1-4419-0465-2_1137
Kyriazis, D. (2015). Antonakis lives on Fokionos Negri, the most beautiful street in Athens. Available at: https://www.lifo.gr/print/geitonia/o-antonakis-zei-sti-fokionos-negri-ton-oraiotero-dromo-tis-athinas
Marinakis, B. (2023). Kypseli: The constant return to the navel of the Athenian world. Available at: https://www.travel.gr/experiences/travel/kypseli-i-diarkis-epistrofi-ston-omfal/
Michani tou chronou (Μηχανή του Χρόνου). Available at: https://www.mixanitouxronou.gr/i-fokionos-negri-otan-itan-rema-poios-itan-o-filergatikos-ypourgos-fokion-negris-pou-edose-to-onoma-tou-stin-plateia-kai-anakalypse-mia-vythismeni-poli/
Mongin, O. (2003). De la ville a la non-ville. In De la ville et du citadin ed. M. Roncayolo et al, Parentheses.
Monumenta. Available at: https://www.monumenta.org/magazine/site%3AR6W6-%3AR6W6-OK7REHR3E/ΑΫΛΗ%20ΜΑΡΤΥΡΙΑ/site%3AOVQ2-H5GRFONKS
Pantaleon, L. (2006, Jan 13). *Isaac Babel, a master of the least, Library/Eleftherotypia*, pp. 16–19.
Pro News Available at: http://www.pronews.gr/istoria/358060_fokionos-negri-deite-ti-metallaxi-toy-astikoy-topioy-stin-athina-foto
Rentzos. G. (2006). *Anthropogeographies of the city, TYPOTHITO, George Dardanos.*
Stefanou, J. (1988). *The urban landscape.*
Stefanou, J. (2017). *Keynote speech at the 7th interuniversity seminar on sustainable development, culture and tradition.* Institute of Sustainable Development, Culture and Tradition, Ano Syros, Syros, June 26–July 1
Technical Chamber of Greece. (1935). The Technical Program of the Municipality of Athens in the magazine. Technical Chronicles, issue 79, 80, 81.
Theodosopoulos, D. (2016). *When Fokionos Negri was a stream//Photo-topographical documentation of the Levidis stream.* Available at: https://geomythiki.blogspot.com/2016/12/blog-post_18.html
Stefanos, T. (2021). *The pale fountains of Fokionos Negri.* Oxy Publisher.
UNESCO (2011). *Recommendation on the Historic Urban Landscape.* Available at https://whc.unesco.org/uploads/activities/documents/activity-638-98.pdf
UNESCO (2016). *Culture Urban Future: Global Report on Culture for Sustainable Urban Development.* Available at: https://unesdoc.unesco.org/ark:/48223/pf000024599
Vaiou, N., & Lafazani, O. (2015). *Kypseli and its Agora: conflict and coexistence in the neighbourhoods of the centre.* Available at: http://www.athenssocialatlas.gr/arthró/%20kypseli/
Vaschenhoven M. (2003, 23 Feb). The genealogy of Kypseli. In *Newspaper Kathimerini, Seven Days, Kypseli. The urban yesterday, the colorful today*, p. 4–7.
Vatopoulos, N. (2002). *The face of Athens.* Potamos Publications.
Webber, A., & Wilson, E. (2008). *Cities in transition: The moving image and the modern metropolis.* Great Britain: Wallflower Press.
Woman Toc Men (2018, Dec 08). Available at: https://www.womantoc.gr/men/article/me-thrylika-klab-me-bon-viver-kai-alites-o-giorgos-malekakis-milaei-gia-ti-fokionos-negri/
Woman Toc Stories (2023, Aug 09). Available at: https://www.womantoc.gr/stories/article/enas-playboy-mia-striper-kai-ena-labero-klab-sti-fokionos-negri-ton-60s/

Chapter 5
Abandoned Factories and Industrial Landscape in Albania

Mariateresa Giammetti and Kreshnik Merxhani

Abstract The article presents the first results of the research jointly developed by the Departments of Architecture of Epoka and Federico II Universities. The study aims to highlight the legacy of the industrial heritage realized by the Albanian Communist Regime. The research hypothesizes that one such study may be useful in defining design proposals for a possible integrative strategy to the development future scenarios defined by the projects and planning tools in force for Albania. The study is searching for architectural complementary ideas that can activate spatial, political, and social dynamics born from the valorization of the industrial archaeological heritage legacy in the processes of urban and territorial transformation. The research aims to have an impact on the protection and valorization of the historical heritage and abandoned spaces, such as disused industrial areas. Its method consists of learning from the historical dynamics of the Albanian urban and territorial program derived from the transformations activated by the Communist Regime, in order to understand the values of the strata that express its historical evolutions. The tool used by the study is a documental analysis, conceived and developed within the research, useful for constructing maps on the interweaving of the socio-anthropological context and the topological characteristics that define the peculiarities of Albanian industrial landscapes.

Keywords Industrial heritage · Third landscape · Pollution · Revitalization · Brownfields

---

M. Giammetti (✉)
Department of Architecture, Federico II University, Naples, Italy
e-mail: mariateresa.giammeti@unina.it

K. Merxhani
Department of Architecture, Epoka University, Tirana, Albania
e-mail: kmerxhani@epoka.edu.al

F. Naselli et al. (eds.), *Self-Sufficient Cities and Sustainable Habitats*, The Urban Book Series, https://doi.org/10.1007/978-3-032-10337-6_5

5.1 Albanian Industrial Landscape

Cities are rapidly changing: the main driver of this process is still the economy that today has abandoned the secondary sector to concentrate on the tertiary one. Industrial areas, despite having exhausted their urban role, are still standing as ruins/ scraps of architecture, as expression of the so called Third Landscape (Clement, 2005). However, despite the close link between economic trends, landscape and architecture, it is important to highlight that the shape of the city and the landscape is a complex structure that is the result of the interaction between human and non-human, between human beings and the environment.

The issue of Third Landscape is particularly acute in Albania. The Communist Regime decided to transform Albania from an agricultural country into an industrialized one. Hoxha initiated an industrialization process modeled according with the major communist powers, China, and the Soviet Union. When Communism fell, privatization was initially attempted, then the factories were permanently decommissioned and largely despoiled. Today, many disused industrial areas have become neglected spaces, drawing a spectral and polluted landscape. These abandoned scenarios add up to Tirana's turbulent and chaotic post-socialist urban development that radically challenged the social and architectural meaning of its identity: on one hand, the architectural historical patterns characterized by pre-socialist and socialist stratifications, and on the other hand, the disproportionate post-socialist built environment that led to a hybrid identity. It seems that new Albanian urban planning and architecture have forgotten these large industrial areas that look like ecological black holes: often the city and the surrounding area grow and transform, as if there were an invisible wall that prevents them from being perceived. However, it is not that: not addressing the issue of abandoned industrial areas is a burden not only on the environment and health of the Albanian people, but also on their cultural and architectural redemption. It would be desirable to keep a constant focus on these areas and their historical memory, to avoid the glitter of skyscrapers, expression of the new neoliberal Albania, definitively overshadowing this great unresolved issue. Starting from this cultural context, the article will describe a historical excursus on the reasons that led to the birth of the so-called Albanian 'industrial landscape' and its propaganda value during the Communist Regime. The article aims to emphasize the importance of knowing the historical–critical and architectural interpretations of the Albanian industrial landscape. This analysis could be useful for the discovery of new design topics that could suggest strategies for transforming the Albanian urban fabric, in some way alternative/additional to the current vision of the Albanian urban and territorial shape that is completely flattened on an architectural style derived from a mannerist deconstructivism already transformed into a postmodern phenomenon. Through the analysis of case studies, the article aims to highlight how the status of contemporary urban design also passes through these neglected places that present themselves to the challenge of urban regeneration in large contemporary metropolises and in so-called emerging countries.

In a tribute to Bennardo Secchi, Vittorio Gregotti (Gregotti, 2016) emphasizes the centrality of emptiness as a topic of research and design for the contemporary city. Quoting Secchi, he writes: "The extraordinary expansion of open space changes the status of the contemporary city, as does the new autonomy of the architectural object. [...] These are perhaps the main causes of the profound uncertainty where urban land and its design finds themself today" (Secchi, 2005). But how can we represent (in the sense of making present to those who inhabit it) the emptiness/open space of the contemporary city? It is often a space made up of waste/neglected places: residual urban spaces, areas created by the disintegration of the city, abandoned areas created by the transfer of production activities where decommissioning processes have been started but never completed, areas that have fallen into disuse because they are contaminated, areas occupied by waste (Giammetti & Rigillo, 2020). Despite having exhausted their active role, the industrial areas still stand as archeological ruins/scraps of architecture. The neglected areas and the ruins/scraps architecture are still part of the urban and territorial metabolism, they make up the so-called Fragile Urban Landscapes or Third Landscape that is an undetermined fragment of the Planetary Garden (Clement, 2005). According to the European Landscape Convention, the landscape represents territory as it is perceived by the population; its character derives from the action of natural and/or human factors and their interrelations (Priore, 2006). It is not just an architectural dimension, but the landscape represents the symbolic form (Kubler, 2002; Panofsky, 2013) of a broader anthropological and ecological dimension that includes the economy, politics, society, in other words, the landscape is closely related to culture in its broadest sense.

From a certain point, Albania's economic history is a paradox: despite the considerable presence of raw materials, mineral resources, fertile land and large hydroelectric capacities, the history of Albania's industrial development and economy was marked by moments of great crisis that led Albania into stories of economic backwardness for several centuries. Even in antiquity for example, Selenica was renowned for bitumen mining; Durres was the port on Via Egnatia, an important transit route from the Mediterranean through the Balkans to Constantinople. Under the Ottoman domination, economic activities were mainly concentrated in the cities of Elbasan, Shkodra and Durres. However, the country remained anchored to feudal agriculture, where fertile land was owned by powerful landowners, predominantly Muslims, who had little interest in modernization or market production, while mountain farmers survived on small plots.

Albania's morphology is characterized by mountain crests that occupy a large part of the country arranged parallel to the coast, between which are deep valleys, where rivers and streams often flow, sometimes forming imposing waterfalls and gorges. This territory is sometimes hostile: the layout of the mountains isolates the valleys and makes communication between the interior of the country and the coast difficult. Here, the landscape is punctuated by small agricultural and pastoral villages with a few small and medium-sized towns. Since Byzantine times, land and privileges in these areas were donated to nobles and functionaries at the detriment of small landowners and tenants. The feudal system combined geographical isolation, due to the particular morphology of the country, with economic and cultural isolation, to the detriment of modernization.

The so-called first industrial revolution, the socio-economic process that transformed agricultural societies into mechanized one, only touched Albania slightly in the second half of the nineteenth century. At that time, Albanian production was largely artisanal and took place in small workshops equipped with basic mechanical devices, located on the ground floor of single-family houses. Production included textiles, felt processing through a mechanism called Derstil, brickmaking, leather shoe production, domestic metalworking, weapons manufacture, and oil production. Until Albania's independence from the Ottoman Empire in 1912–13, the country developed only limited mechanical industrialization. When it became independent from the Ottoman Empire, Albania was a country with a mainly agricultural and backward economy, lacking road and rail infrastructure and with widespread illiteracy. After the First World War, prompted by several international agreements, many Italian companies established new factories in Albania and the Azienda Italiana Petroli Albania (AIPA), a company active from 1925 to 1943 and controlled by Italian State Railways and AGIP, began the extraction of oil. However, the Albanian economy continued to be weak: the agrarian reform, prompted by Ahmet Zogu, failed. The farmers, still mainly equipped with wooden ploaghese, could not meet the food needs of the growing population. The mining of coal, copper and chromite gradually was under the Italian companies control and the Albanian economy became increasingly dependent on Italy. Since the independence from the Ottoman Empire to just after the First World War, Albania went through an economically and politically unstable period. Industry continued to be mainly supported by foreign investments, particularly by Italian companies. Many factories were dedicated to processing Albanian agricultural products and to mechanizing some handicraft sectors, particularly textiles. Albania became part of the Italian Fascist Empire that concentrated its investments mainly in the central and southern plains. After the end of World War II and the fall of Fascism, Albania became the Socialist People's Republic of Albania, which passed first through Socialism, then Communism and finally to a dictatorial form of collectivism that was mainly ruled by Enver Hoxha. After the proclamation of the Socialist People's Republic of Albania in 1944, the government sought economic support from Yugoslavia and adopted an economic model derived from the Soviet Union. The Albanian government nationalized land ownership and companies and massively expanded public heavy industry.

After the Second World War, Albania entered into a new phase of industrialization with an intensity that was unlike anything that had gone before. Many industrial buildings were severely damaged by the war and needed to be rebuilt or were demolished. Any facilities or sites that were owned by foreign companies were nationalized by the Communist Regime. This process was completed by 1947. By which time Albania had halted the pre-war trajectory of industrial investment and development by foreign companies. A series of five-year economic plans based on the Soviet model were devised which emphasized communist industrial advancement as the essential component for future progress.

The railway network began with the first connection between Durres and Peqin in 1947, but the lines were mainly built by unskilled personnel and based on volunteer work missions with railway equipment that was often second-hand. After the political crisis between Yugoslavia and the Soviet Union in 1948, Albania approached Moscow for financial support and technical assistance to develop infrastructure projects. The Soviet Union was an important buyer of Albanian raw materials. However, the agricultural sector remained neglected and Albania was forced to import wheat from Russia and accumulate debts. However, Enver Hoxha did not want entirely to surrender himself to Soviet Union. Since 1961, he formed an alliance with China in order to further advance the project of developing an Albanian heavy industry able to process some of the locally extracted raw materials: copper in the wire and cable factory in Shkodra and oil in one of the first Albanian refineries in Fier; nickel-infused iron from the Pogradec region mainly destined for the steel industry in Elbasan. In 1970, the connection of the entire country to the electricity grid was announced. The 1970s probably represented a decade when industrial production exceeded the agricultural one. However, industrial production was heavily dependent on economic aid from China. When China cut off financial support in 1978, Hoxha declared that Albania would reject all foreign aid and proclaimed his vision of a self-sufficient country. He started a process of economic collapse that peaked in the 1990s with looting and political unrest.

5.2 The Communist Heritage: The Industrial Landscape in Socialist People's Republic of Albania

The article describes a historical analysis concerning the industrialization processes that have affected Albania from the late Nineteenth century to the present and proposes a reading of some examples of industrial sites born in those period that are now disused. In parallel, the article intends to demonstrate how the statute of contemporary urban and landscape design also pass through these disused places: abandoned industrial areas that represent one of the greatest future challenges of urban regeneration in the big cities and so-called emerging countries. Beginning with a historical-architectural analysis, the article aims to demonstrate how abandoned factories areas and their industrial landscapes can have new meaning for Albania's contemporary urban palimpsest and spatial metabolism. They could be "nutrients" aimed at: (a) being reused as public spaces and facilities; (b) creating a widespread system of re-naturalized free soils in the dense city.

Fig. 5.1 Textile Kombinati Stalin, historical images, and current state

5.2.1 *Textile Kombinati Stalin, Tirana*

The design and construction of the ‘Stalin’ textile factory dates back to the early 1950s. It is located southwest of Tirana, in an area that was originally swampland and was reclaimed for the construction of the factory that produced fabrics and clothes in cotton and other materials, as well as tapestry fabrics. There was also a specialized production line for the manufacture of military underwear and cotton gauze used for medical needs. This manufacturing complex became so radically part of the history and identity of the city, that it gave the name “Kombinat” to the neighborhood of Tirana where it was built (Fig. 5.1). The main entrance was designed to be monumental, opening to a square with a statue of Stalin in the middle. The factory began to go to ruin in the 90 s, when the machinery was looted and sold as scrap. Since then, many of the buildings have been destroyed and, currently, most of the structures are in a serious state of decay. Part of the complex has been squatted by families and today it is estimated that around 700 people live there in very precarious conditions. The factory is now abandoned and looks like an empty shell bubbling with poverty and unemployment.

5.2.2 *Tobacco Manufacturing Stamles Durres*

Albania is well known for the excellent quality of its oriental tobacco. The cigarette industry in Albania began in 1924 with the foundation of “Stalme” cigarette factory in Durres. In total, 16 cigarette factories were opened for the national market in the

Fig. 5.2 Tobacco manufacturing Stamles Durres, sequence of images from the foundation in the early XX century, to the disused phase and to the nowadays new building realized in substitution of the hold factory

20s, 30s, and 40s. In 1947, all the factories in Albania were nationalized, most of them were closed, while the larger ones, such as Stalmes, were renamed and modernized. During the Communist Regime in Albania there were four cigarette factories: the Tirana factory, known as "Flora," which was closed in the 60s, the Durres factory, the Shkoder factory and the Gjirokastra factory. The factories in Tirana and Gjirokastra produced cigarettes for the national market, while the factories in Durrës and Shkodër produced cigarettes for both the national market and for exportations. During the 1960s, Albania began exporting cigarettes to all the countries of Eastern Europe, in particular Hungary, Poland, Czechoslovakia, and the German Democratic Republic. In the 1980s, most of the cigarettes produced in Albania were exported as they were a good source of foreign currency. In the 1980s, the Durres Cigarette Factory started producing several brands under license from Bulgartabac (Arda and Steëardess) and signed an agreement with the Polish Cigarette Monopoly to produce "Dukat" cigarettes, made with Polish duahn, to be exported exclusively to Poland (Fig. 5.2).

After the fall of the Communist Regime and the advent of economic liberalism, Tobacco Manufacturing Stalmes was almost completely demolished and today the Albanian College Durrës, an enormous neoclassical building, stands on the site.

5.2.3 *Petrolia*

The first oil exploration in Albania began before the First World War. The Italian State Railways were among the first Italian concessionaires, in 1925 with 46,000 hectares, to which another 117,000 hectares were added in 1926. Starting from these concessions was born AIPA, Azienda Italiana Petroli Albania, also known as Azienda Italiana dei Petroli Albanesi. It was a company controlled by Ferrovie dello Stato and AGIP, active from 1925 to 1943. The valley of the river Davoli supplied 10% of Italy's annual oil requirements. In a few years, the small rural village of Kucova was transformed into a real industrial city. Kilometers of roads, bridges, oil pipelines, railway stations, oil extraction plants, a small refinery to process crude oil on site, and a military airport were built. The small town was populated by Italian and Albanian workers and employees. During the fascist occupation, Kucova was even renamed Petrolia, city of oil.

Most of the oil was transported to Bari where it was processed and used nationally. After the Second World War, Kucova ceased to be Petrolia and became *Qyteti Stalin*, the city of Stalin. Its industrial activity continued: once the Italians from Aipa had been kicked out, the oil extraction factories passed into the hands of the regime. The city took on the gloomy appearance of a military district, closed off like Hoxha's communist Albania. When the Communist Regime fell in the 1990s, the city began a slow decline, and almost all the facilities closed. Walking through the streets of Kucova today is like taking a long stride through the history of Albania. A history of contested occupations, betrayed ideals, unfulfilled hopes. All this is there among the ruins of what was once one of the most important industrial cities in the country. All around is disaster. Flocks of sheep graze on oil-soiled grass. Oil pipelines, worn by time and neglect, break up the city's roads and pavements. The chimneys of the old extraction plant stand out in the center of Kucova, as imposing as the sense of abandonment they convey.

5.2.4 Elbasan

After the withdrawal of Soviet technical and material aid, due to an international incident in the communist movement known as the Sino-Soviet split, Enver Hoxha chose to bring Albania into the political sphere of the People's Republic of China. China helped Albania develop an even more radical economy based on heavy industry, in the spirit of Stalinist industrial philosophy. China financed the construction and/or expansion of new factories with very high-cost investments such as the Elbasan ferrochrome factory. When Albania was a partner of the Soviets, Elbasan was a minor industrial center; however, thanks to Chinese funding and technical assistance for the construction of the Metallurgical Combine, it became the heart of Albania's black industry. The factory was named "Çeliku i Partisë" and was emphatically referred to by Enver Hoxha as 'Albania's second national liberation' (Fig. 5.3).

Placed in the valley of the River Shkumbin, near the town of Elbasan, the Metallurgical Combine became the pride of communist achievement in Albania. Construction of this massive diff industrial complex began on the 8th of November 1971, which was the anniversary of the creation of the Albanian Communist Party hence the name given to the factory. This was an Albanian–Chinese cooperation initiative and all manufacturing projects were overseen by experts from China, who assisted for about 8 years. The complex employed over 10,000 workers over a number of different enterprises such as the Nickel–Cobalt Plant, Cast-Iron Plant, Steel Plant, Carbonite Production Facilities, Refractory, Brick Factory, Thermo-Electric Plant

Fig. 5.3 Elbasan metallurgical combine, images of the current state that express both its character as a third landscape and the high level of pollution

(TEC), Oxygen Plant, Chemical Plant, Supply Enterprise, Enterprise of Transportation, Institute of Design and Metallurgical Studies, sourcing water for lime processing, and many other jobs (Fig. 5.4).

5.3 Conclusions

Disused industrial areas are the result of transformation processes induced by lines of economic development that have exhausted their urban and social function. However, despite the close link between economic trends and the urban landscape, it is good to keep in mind that the shape of the city and the landscape is a complex structure that is the result of the interaction between human beings and the environment. It is not only an economic dimension, but also a social and cultural one. Neglected areas (abandoned buildings and factories, uncultivated land, rubbish dumps, etc.) are also part of the landscape. In particular, they constitute the so-called fragile landscapes or Third Landscape or a new idea of industrial landscape that can be understood as an indeterminate fragment of the Planetary Garden (Clement, 2005). The issue of abandoned factories is particularly acute in Albania. The Communist Regime decided to transform Albania from an agricultural country into an industrialized one. Hoxha

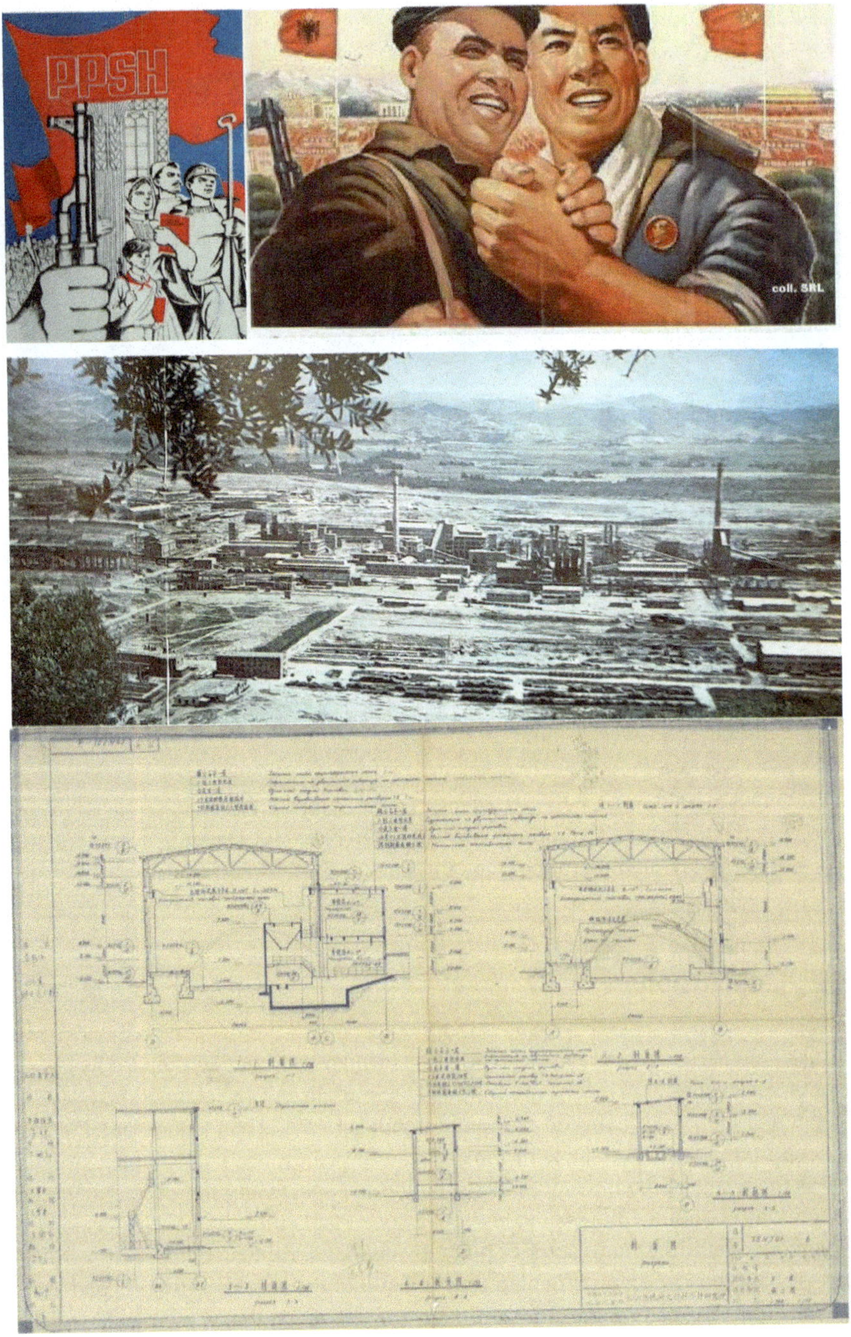

Fig. 5.4 Collage representing the storytelling linked to Elbasan metallurgical combine. Top right, images from journals expressing the spirit of communist propaganda linked to the industrial development of Albania and (top left) the political alliance with China; executive drawings of the factory with Chinese writing (middle), indicative of the support offered by China for its construction and (down) image of the Plant in 1980. *Source* V. Koçi

started a process of industrialization based on the model of the main communist countries.

After the fall of the Communist Regime, the factories were definitively decommissioned and largely ransacked. Today many disused factories have become abandoned spaces, creating a ghostly and polluted landscape in Albania. It seems that the new Albanian urban and architectural planning has forgotten these large urban and ecological black holes. Facing the problem of abandoned industrial areas is a problem not only for the environment and the health of Albania, but also for its cultural and architectural redemption.

Albanian urban and regional development still does not pay adequate attention to the intrinsic historical values expressed by the architecture built over the last one hundred and fifty years of Albanian political history. Probably, it is due to ideological reasons that lead to the historical removal of the Communist Regime, but this is resulting in the loss of the legacy of many other components of the Albanian cultural identity. During Communism, propaganda was entrusted to the architectural topic of the factory and, consequently, to the economic topic of industrial development. Today, the propaganda used by the Neoliberal Capitalist Regime in power is the "official city" that appears along the main avenues and axes with its skyscrapers and disproportionately sized block buildings and their "pixelated style" facades. In both cases, history is a troublesome witness that should not be ignored. Excluding a few redevelopment projects, once again in line with the character that has typified the cultural flattening of the periods of political oppression in Albanian history, it seems that the attention is focused on the propaganda aspects of which architecture can be a vehicle.

The research on the Albanian industrial landscape, a legacy of the lack of economic impetus that the Communist Regime tried to give to the country, aims to provide an updated diagnosis of what Albania was like in the past and what it has become today. The case studies have demonstrated the multiple dimensions of the landscape, of the architectural heritage and of the urban fabric: the physical space, social life, human and political interactions and the framework of historical and economic processes.

As demonstrated by the results it is obtaining, this research work will be able to provide recommendations and practical suggestions for a systemic regeneration of the landscape and cities that considers the legacy of Albanian political and cultural history as a tool to promote an effective improvement in the quality of real life 'beyond' the current formal urban process in progress.

The recommendations that can be summarized here are: the analytical documentation collected for each abandoned industrial area must include the buildings but also the environmental conditions of the context and data relating to its pollution; the population must be included in the re-evaluation process; a multidisciplinary approach is necessary to understand the complexity of the site; regeneration projects for disused industrial areas must include long-term objectives for their reclamation and short-term solutions for their immediate temporary reuse.

Acknowledgements This chapter presents the research achievements developed jointly by the authors. However, Kreshnik Merxhani's contribution primarily focused on parts 5.2, while Mariateresa Giammetti's contribution specifically concentrated on parts 5.2 and the conclusions. Both authors drafted the introduction.

References

Clement, G. (2005). In F. De Pieri (Ed.), *Manifesto del Terzo paesaggio*. Quodlibet.
Giammetti, M., & Rigillo, M. (2020). *Re_Thinking urban mining*. In F. Moccia, & M. Sepe (Eds.), *Urbanistica ed informazioni n. 289 s.i./2020*. Inu Edizioni, p.114.
V. Gregotti. (2016). https://www.casadellacultura.it/421/bernardo-secchi.
Kubler, G. (2002). *La forma del tempo. Storia dell'arte e storia delle cose*. Einaudi.
Panofsky, E. (2013). *La prospettiva come forma simbolica*. Abscondita.
Priore, R. (2006). *Convenzione europea del paesaggio, il testo tradotto e commentato* (pp. 2–3). IRITI Editore.
Secchi, B. (2005). *La città del Ventesimo secolo* (pp. 185–186). Laterza.

Chapter 6
Tirana-Next: Exploring Potential Strategies for the City's Sustainable Third Landscape

Fabio Naselli and Mariateresa Giammettii

Abstract The article aims to present the first research results jointly developed by the Departments of Architecture of Epoka and Federico II Universities. The research aims to set design proposals for a possible "missing" strategy on the horizon of future developments envisaged by projects and planning tools in force for Tirana. These are complementary strategies that may activate spatial, political and social dynamics born from the real involvement of citizens in the urban transformative processes. The research intends to have impacts aimed at protecting and enhancing the legacy of diffuse historical heritage and waste-spaces that contribute to realizing the horizons of Tirana's Third Landscape. The research method is to learn from the evolutionary historical dynamics of Tirana's urban palimpsest in order to understand the values of the layers that express their historical transformations. The tool the study uses is a multi-criteria analysis, devised and developed within the research, which is useful for building maps on the interweaving between the socio-anthropological context and the topological characters that define the peculiarities of Tirana's urban landscapes.

Keywords Tirana-next · Third landscape · Urban strategies · Sustainable city landscapes

6.1 Introduction

The article presents a summary description of the studies and investigations regarding the research program jointly developed at the Departments of Architecture of Epoka and Federico II Universities, and in conclusion, discusses the first results of the research project.

F. Naselli (✉)
Architecture Department, Epoka University, Tirana, Albania
e-mail: fnaselli@epoka.edu.al

M. Giammettii
Architecture Department, Federico II University, Naples, Italy
e-mail: mariateresa.giammeti@unina.it

F. Naselli et al. (eds.), *Self-Sufficient Cities and Sustainable Habitats*, The Urban Book Series, https://doi.org/10.1007/978-3-032-10337-6_6

The research started in 2019 as a master's thesis class of some students of the Department of Architecture at Epoka University which aimed to provide an up-to-date diagnosis based on the analysis of Tirana: how it was in the past, how it is now, how it is changing, which are its current conditions, what is being carried out by the current Master Plan "Tirana 2030", that came into force in 2018 and the other macro-projects ongoing to transform the city," into a "Modern City." Tirana-Next, this is the name of the research, an open pilot research and, precisely due to its own character, it has been improved with the joining of the research working group from Federico II University. The partnership between the Departments of Architecture of Epoka and Federico II University has meant implementing a number of case study areas and developing an additional channel of analysis on the topic of "fragile" and "waste." The research on Tirana's historic architectural heritage and its characters was framed within the broader horizon of fragile landscapes and declined as waste-landscapes. Consequently, the analysis was expanded to a wider range of spaces that can be categorized as waste spaces. Fragile is what appears particularly vulnerable and requires special attention to preserve its beauty and integrity, since it has great sensitivity to human impacts. Waste is what is abandoned, left in a state of decay due to changes in human activities. These two topics offer an interesting key when cross-referenced with the Master Plan "Tirana 2030," where skyscrapers and out-of-scale buildings lie or overlap, heedlessly, with fragments of the ancient and recent historic texture. Tirana's old palimpsests are downgraded from an urban matrix to abandoned or wasted space to be demolished to reclaim soil and build again. In the future city, the empty spaces, understood as spaces in between, fragments of the third landscape born from the frenetic growth of the informal city, rather than being an opportunity to generate useful breaks in the dense city (equipped green spaces for communities, informal markets strongly rooted in local culture, etc.), become yet another opportunity of real estate speculation. In its renewal efforts, Tirana exhibits the brutality and carelessness of the advancing new and the fragility of the traces of its past. Tirana leaves the layers of its historical urban palimpsest in a state of neglect, to the category of waste and phagocytes them into the speculative processes of "modernization." Starting from these observations, the research aims to identify regenerative strategies that are "missing" in the Master Plan in force. The research is oriented to investigate urban development strategies that consider the ancient and recent historical dynamics of Tirana's urban palimpsest and the values expressed by the layers that explicate its historical evolution. Through multi-criteria analysis, the study is oriented to collect data to define useful design proposals for reinserting the diffuse historical heritage and the waste spaces into the active chain of urban metabolism. The research aims to validate and define a proposal for at least two possible "missing" strategies: (A) to protect and enhance the traces of Tirana's diffuse historical architectural heritage; (B) to enhance the system of waste-spaces that contribute to defining Tirana's "Third Landscapes" system.

6.2 Analysis and Thematic Horizons

The research aims to provide criteria for developing an updated analysis on the urban palimpsest, architectural cultural heritage and landscape. Starting from the analysis outcomes, the study develops design criteria oriented to activate new forms/actions of urban metabolism. The analysis is developed from two thematic horizons:

- How Tirana's physical and socio-cultural space was in the past and how it is becoming today.
- What the "Tirana 2030" (Bulleri, 2018) Master Plan and other ongoing macro-projects are and will be put in place to transform Tirana into a "modern city" and how they will act on its social structure.

Through articulated surveys within Tirana's neighborhoods, researchers are exploring possibilities to implement documentary research by sharing with city residents the multiple experiences related to the values produced by the city's physical, cultural, and political space and the changing relationships of humanity and spatiality that surround them. The investigations are exploring physical space (solid landscape), social life in the streets (soft landscape), human interactions (human landscape), virtual spaces beyond the physical (digital landscape), result spaces and drosscape (third landscape), and the process frame (cycles and process landscapes). As demonstrated by the early results, it is expected that this research can provide recommendations and practical suggestions by investigating and analyzing the links between citizens and these elements. Its goal is to discover more appropriate systemic regeneration actions useful to the real quality of life in Tirana's neighborhoods, which also go beyond the current formal urban process.

Cities like Tirana, in a phase of transition, are changing very quickly. Tirana's turbulent and chaotic post-socialist urban development has radically challenged its identity: on one hand, the city's historical urban pattern characterized by pre-socialist and socialist stratifications, on the other, the disproportionate post-socialist-built environment with its neoliberal character that has led to a hybrid and cacophonous identity, which demands "Quo Vadis" Tirana?

Tirana's traditional small-scale houses are replaced by large urban blocks with minimal public and semi-public spaces and facilities. This has led to a dramatic increase in urban density, which has become one of the highest among European cities. The skyscrapers and out-of-scale buildings lie or overlap, heedlessly, with fragments of the historic architecture, which is downgraded from a matrix of the urban palimpsest to abandoned or waste space to be demolished in order to reclaim land and build again. The principal criterion of urban transformation is still only the liberalist economy, as in the last century. Urban metabolism produces waste that corresponds to parts of the city itself. Whether they are parts of the historical texture or portions of unbuilt areas, leftover from the informal urban growth mechanisms activated immediately after the end of the socialist regime, or the result of current transformation processes, all these urban-waste areas combine to set the Third Landscape of Tirana. This process is like a litmus test of all cities affected by urbanism, mainly inspired by the neoliberal economy. Within the urban textures of such a Tirana, the Third Landscape (Clement, 2005) still stands as a ruin/waste of architecture. All around, the new city grows and consumes new soil and creates imbalances that can no longer be ignored.

6.3 History and Layers

As is generally the case with Mediterranean cities (Naselli, 2015), the urban palimpsest consists of the overlapping of many historical and cultural layers. Specifically, in Tirana, there are pre-Ottoman settlements, the Islamic-Ottoman city, the modern city (promoted by King Zogu I and developed by the Italians), the socialist city, with its residential blocks and new *extra-moenia* work settlements and the informal neoliberal city of the immediate post-regime period. These layers add up to the turbulent post-socialist process of reclaiming private property and to the chaotic contemporary city, which moves according to the slogan of the 'modernization' (Aliaj & Lulo, 2003) required to completely leave behind the legacy of the communist regime.

Layers surface among the massive and impactful volumes of contemporary Tirana: through a curious perceptive process, they take on an informal urban character. This is a spontaneous process that has challenged, and still challenges, the deeper existential meaning of the city's urban multi-identity. It presents itself with its familiar human-scale "interior" cityscapes, composed of single-family houses with gardens (both pre-Ottoman and Ottoman), small-scale buildings (rationalist buildings and the so-called Italian villas), socialist blocks and a vast array of informal micro-realizations and spontaneous creative adaptations that make the inner areas of the city deeply representative of a substantial part of local culture and social and family relations (Mali, 2018).

Shaping a new and modern Tirana. This is the objective that the current Albanian political leadership is pursuing. All strategies are acceptable, however, they seem to have overlooked and hidden in the superblocks of the formal city the features of the third landscape that characterized Tirana just after the communist period, which has been interpreted only as "chaos." According to the new strategic lines of "TR-030," private initiative is tasked with providing a new "order" to the existing "chaos." Building "bombings" and urban replacements are the tools used to make the "new order." Parts of Tirana, made by the overlapping layers of the historic city and the "informal spontaneity" generated immediately after the end of the communist regime, first become waste-spaces and then are demolished to free up land for new buildings. The first results of the analysis referred to in this article show that, despite the merits and innovative ideas outlined in the current urban planning tool, something is still missing in the development strategies for the future of Tirana. In particular, the analysis revealed that the "order," although planned, remains imprecise concerning the deep structures that underline the symbolic form of the city of Tirana. It can therefore be deduced that greater protection should be introduced for the "chaos" of the informal city, which is, after all, one of Tirana's distinctive characteristics.

In 2022, Tirana had 906.166 inhabitants shared in an area of approximately 1.110 km^2. The city collects more than 30% of the entire Albanian population (2.793.592) (https://www.instat.gov.al). Tirana is located in a wide alluvial plain positioned in the center of Albania.

The city is surrounded by hills that divide it from the Adriatic coast and mountains that separate it from the country's hinterland. The city is crossed in an east–west direction by two rivers: Lana River, which runs through it from side to side, and Tirana River, which bounds it to the north. Within the city, there are also two lakes, Lake Farka and the so-called artificial lake, a point of interest for sports, recreation and leisure activities. At present, the structure of the urban plant is articulated in blocks and super-isolates surrounded and configured by the main road axes (Boulevards) and three main inner rings. Within these city blocks (superblocks), bound by socialist or new post-socialist buildings, is contained the informal scenario of the real city's housing identity. Beyond the border of the superblocks, one can discover pieces of the city from the Ottoman period, with the characteristic single-level courtyard houses surrounded by white fences with the old wooden gates and on whose roofs one can see prehensile gardens; pieces of the city that still reflect the agricultural character of Tirana's past; buildings stained by the self-building interventions of the communist city, organized in courtyards around green spaces, sometimes still preserved, sometimes occupied by the dehors of the recently opened cafeterias; ground floors that expand toward the street and house commercial activities based on neighborhood micro-economics; irregular pieces of the urban system that turn on themselves and whose geometries are the result, sometimes of the old Ottoman layout, sometimes of the processes of land restitution to owners following the end of the collectivist regime during which the socialist city had developed heedless of the original properties.

It is not possible to understand the social and spatial potential of neighborhoods if one looks at them only at the large scales usually used in urban planning. On the contrary, it is necessary to frame the neighborhood at the intermediate scale proper to urban design in order to understand its values and the uniqueness it expresses from a historical architectural point of view. Even more so, it is necessary to explore neighborhoods not only through maps, but by physically going to these places and inhabiting their spaces, exploring their streets, alleys, the richness of their systems made of fragments of history and spontaneous and informal adaptations to sudden new freedoms, the multiplicity of styles and transformations, as well as to the variety and density of social relations and interrelationships. In this race to modernize Tirana, it becomes urgent to turn the spotlight on the neighborhoods and waste spaces, to consider them in a more meaningful way than the current TR-030 urban plan does, and not to lose the system of values they represent. These are parts that need regenerative attention that consider the complexity of elements and factors. It is an action that must go hand in hand with the reconnaissance on the one hand of the drosscape and waste spaces and the other hand of the still existing historical heritage. They are in danger of destruction because they are not recognized in their uniqueness as a monument nor as a fragment of a more complex historical and/or cultural system where it is also necessary with the recognition of the value of transformations and spontaneous settlements, adaptations that have arisen from the readjustment of spaces, volumes and living activities.

The origins of the first settlement can be traced back to the third century AD, although the region that corresponds to its territory today has certainly been inhabited continuously since the Iron Age. The area was populated by the Illyrians and was probably an agricultural and commercial center in Roman and Byzantine times as well. During the Ottoman Empire, the settlement acquired the status of a city: essential facilities, such as the market, hotel, hammam, mosque, and bakery were built (Aliaj & Lulo, 2003). It is possible to identify two periods when the city alternated between organic and informal growth and more regular and rational transformation: the Ottoman and pre-Ottoman periods alternating with the settlement of King Zogu the first and, in continuity, the subsequent Italian colonial period; the socialist period alternating with the end of the communist regime until the current "restoration" phase that began in 2005. These events shaped the footprint of the present city. With independence, Zogu the First initiated a radical redesign of the city, which by 1920 was also proclaimed the nation's capital. The informal city gave way to the first approaches of a new formal city: indications and decisions came from the Albanian political center, while the design proposals came from European specialists. Italian architects played a great role in this phase and, even before Mussolini's invasion, had the privilege of being able to start the process of reconfiguring the city through the creation of the first long boulevard. It was a long and wide road axis that cut the city in two and laid the basis for the configuration of the new urban form that would give rise to the current reticular footprint structured in radial axes and ring roads, overlapping the pre-existing organic parts of the city. With the end of the communist regime (1994) and the establishment of a new democratic republic, still immature and weakly represented, a new season of "informality" began. The new democratic system allowed the possibility for "everyone" to implement their own choices: it allowed to carry out building of transformations and urban structures that had been desired and hoped for years. Democracy reopened the question of the physical configuration of the city. The years immediately after the end of communism can be considered a long phase of self-determination, privately or collectively, that pervaded not only Tirana but all of Albania.

The city changed again by indulging in "organic and informal" choices by seeking and giving answers to minute, everyday needs in a true bottom-up process, by pursuing demands and needs finally freely expressed and realizable. Volumes were changed, uses were changed-especially on the ground floors. New buildings were hooked up with creative interlocks, and new houses and new buildings were built in the voids once intended for the collective spaces of regime architecture. The latest transformative phase began in 2005 with the first stable and recognized democratic government, which restored some order and a new democratic respect for rules. Despite this richness and overlapping of successive layers in the configuration of the city, Tirana continues to be considered a 100-year-old city, erroneously starting from 1920, contributing to accentuating the oblivion of the oldest parts of the city.

6.4 Methodological Criteria

If the social sphere is one of the most important factors in the urban environment, the interactions and relationships between non-human and human beings increase the relevance of common spaces and their spatial needs. It is well known that a high degree of quality of life and well-being of people and the natural environment directly affects the improvement of the spatial quality of public urban spaces and landscape (Pashako, 2016). The public sphere of the city, streets, squares, parks, etc., is like an active stage, a catalyst for social interaction that marks the peculiar dynamism of the city. To understand the declinations Tirana offers of this dynamism, it was decided to go even deeper into the analytical phase of the research.

The scale of investigation was focused on Tirana's suburban centralities and neighborhoods: here we can better understand not only the public sphere of the city (streets, squares, parks, etc.), but also the topographical relations between public spaces and private (residence) and semi-private (work) spheres. It is a complex multi-reality, composed of mixed parts of the city, where space has been studied not only in its physical dimension, but also in its socio-cultural dimension and the relations between both. The neighborhood is the basic unit of the community's system of human and social relations, at the local scale as well as the global scale. For this reason, the scale of analysis and understanding of case studies was the neighborhood, central or peripheral. The superblock, or a homogeneous sub-part of it, was fixed as a tool to study the city center of Tirana. For peripheral centralities, the tool was the small urban agglutinations that have arisen because of the development of the informal city. Here, the redevelopment of low-quality places, often forgotten or simply ignored by dominant growth processes, can really enable a true increase in local urban quality. Small spaces that, if they are read as a system of linked public or collective micro-spaces, can interconnect with a network of pedestrian paths and generate an alternative mobility system (bicycle or electric vehicles). Safe spaces, walkable and fully inhabited by neighborhood life. Short paths, walkable in 15 min and interspersed with spaces for aggregation, exchange, and socialization. They are suitable for safe access to all places, and their activities meet daily and weekly needs and return priority to the pedestrian in the use of urban spaces for normal daily activities. Tirana's neighborhoods, especially those hidden in the superblocks, are the places where the system of urban spaces appears more collective than public. From this point of view, they are more closely connected to real life and the system of punctual relations that mark the dynamics of these inner parts. They are topologically very different from the publicized image of the future Tirana and its skyscrapers. Parallel to the neighborhoods contained in the superblocks, as soon as one moves from the city center to the innumerable peripheral centralities that make up Tirana's urban structure, it is possible to find reinterpreted or reinterpretable waste spaces as a shared space to support the activities of citizens in the neighborhood. These are spaces of different sizes: from the uncultivated and unpaved roadsides where small food markets take place every morning, animated by farmers from the countryside near the city to sell their products; empty spaces discarded by urban growth where there are informal markets, makeshift housing, play spaces; large socialist manufacturing buildings now disused, around which small settlements of workers had grown up and which now lie like great skeletons squatted by people who have turned them into residences or reused them again as places of production. To deeply study this complex system of topological relations, the process of investigation and analysis was conducted on two parallel tracks: at the urban scale and the neighborhood scale, although the neighborhood scale played a predominant role. The analysis started with recomposing and updating the knowledge framework of the places. One of the first stages of the investigation was conducted precisely on the structure and contents of the planning instruments in force, retrieved both through the documents released by the public administration and through the official websites of the city of Tirana and the National Agency for Spatial Planning (AKPT), as well as reported on the website of the Boeri Studio, consultant of the Tirana Urban Plan. It became clear that there was a need to implement the data with information regarding people and their activities in order to collect and manage the wide variety and variability of data and information from a multicriteria survey. The investigation was developed in the field to analyze in depth the topological relationships between urban and architectural space, people and systems of material and immaterial values linked to the *modus habitandi* of the citizens. The analysis was conducted on portions of the city capable of significantly expressing the most deeply rooted symbolic forms of living in Tirana and the contradictions to which the city subjects them today. For this reason, the neighborhoods within the superblocks and the waste-spaces that are part of Tirana's third landscape were chosen as areas for analysis. The analysis was developed following a complex and multicriteria methodological criterion of work, integrated and measured especially for these types of areas. Regarding superblocks and neighborhoods, the questions investigated were: (1) How is it possible to record the real experience of citizens and understand what services and activities they use "daily" to serve their real lives? (2) How can we work on stabilizing the spatial and socio-economic conditions of the neighborhood through shared strategies and interventions, including punctual and tactical ones? (3) What are the key issues for Tirana's next development? The data collection process supporting the multi-criteria methodological framework has been implemented through an open-source web application that allows scholars, professionals, and ordinary people to create, customize, and share personal maps of spaces that meet the two categories of "discarded" and "fragile." The digital application offers the possibility of plotting routes, inserting points of interest, adding text, images and videos, creating areas and boundary lines and more.

Users can create maps to share data and information and support the knowledge and dissemination of Tirana's hidden heritage. The web application aims both to collect data through the contribution of people not directly involved in the research and to disseminate data and study results through an open-source web tool that contains maps accessible from any device connected to the Internet. The data collected and derived were subsequently analyzed and interpreted and demonstrated how the chosen case studies can be considered paradigmatic for understanding the entire city of Tirana. The bottom-up approach has suggested possible urban planning and design strategies (to be defined and adapted on a case-by-case basis), complementary to official urban planning, that are more localized/modeled on the specific character of Tirana's places and their values. This different approach, which can be achieved by encouraging the full participation of local communities, has as its main objective the promotion of a future Tirana as a city that is closer to its local communities (Naselli, 2015; Naselli & Trapani, 2020). The results of the research focus on the human dimension of urban space, on the design of a city that is safe for pedestrians, livable in its open spaces, sustainable in the relationship between human and non-human and able to preserve the social and relational characteristics derived from the long and troubled history that today configures a city with parts, all different and overlapping, but all present and alive.

6.5 Study Areas and Application of Analytical Methodology

An interactive map was developed through an open-source web app. The map aims to implement data useful to support the multi-criteria survey (Fig. 6.1). The map was activated by researchers, but it is structured to be also opened to other scholars, professionals, and ordinary people. The map is a graphical representation of different variables and criteria used to evaluate and compare different areas that meet the two categories of "waste" and "fragile." This type of map allows complex and multidimensional information to be displayed clearly and concisely. It highlights study areas such as points of interests (POIs) to which it attaches digital documents (documentary material, photographs, videos, plans, interviews, audio files, etc.).

The map uses symbols, colors or gradations to represent different levels or values relating to the criteria considered. For example, different shades of color can be used to indicate the intensity of a particular criterion in a specific area. The map can be combined with spatial analysis tools to obtain an overall assessment or ranking of places according to the different criteria considered. This makes it easier to identify and compare the areas that best meet a series of specific requirements: the objective is to provide a concise and comprehensible overview of a complex set of information and allow a more accurate geographical evaluation that supports multi-criteria analysis and consequent design strategies.

The map can be implemented: it currently contains both areas belonging to the waste-space class and places representative of Tirana's historical heritage. For the research, a selected number of areas were chosen and subjected to study, investigation and analysis. Some case studies are part of the group of peripheral centralities, others are located within superblocks. Specifically, the definition of peripheral centralities refers to some neighborhoods of Tirana which, although not located in the central core of the city, still play an important role in urban and social life and are significant places for the people who live there. One of the main characteristics of peripheral centralities is the presence of commercial activities or public services, hospitals, universities, and active places in the urban metabolism. These types of urban catalysts are often discarded spaces that, although they have lost their active role in the urban

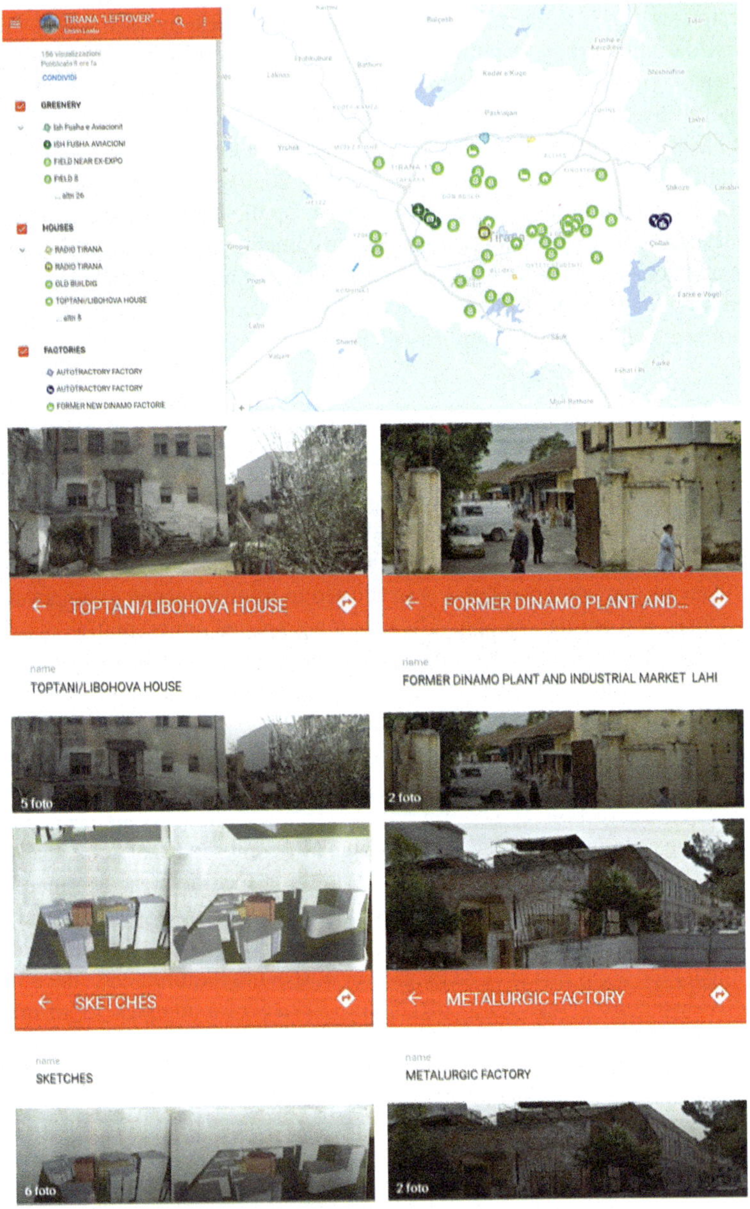

Fig. 6.1 Screen of the multi-criteria and implementable digital mapping interface developed for research

metabolism, have been reinterpreted by the local population who use them to create opportunities for common spaces dedicated to the inhabitants. This phenomenon can be defined as a peculiar urban characteristic of Tirana. However, Tirana's ability to regenerate itself through bottom-up processes that reuse and transform existing heritage, even if in a very precarious way, represents a great resource for providing the city of the future with common goods and livable spaces. The reasons for these spontaneous phenomena of urban transformation are mainly due to Tirana's troubled political history, which has generated many unused spaces that can be reused.

By addressing the challenge of developing peripheral centralities, Tirana could reduce congestion and traffic in its central areas and improve the distribution of services and travel. The development of peripheral centralities could help create a balanced provision of services and opportunities throughout the city, reducing the socioeconomic imbalances between central and peripheral neighborhoods. The research chose as case studies representative of the category centrality peripheries "Uzina Autotraktora," a disused tractor factory dating back to the socialist period that is located east of the center of Tirana, close to the internal mountain system, and the major road axis "Fusha e Aviacionit," originally an landing strip, that is located to the west of the center between the two surrounding ring roads (Fig. 6.2).

Uzina Autotraktora is an abandoned medium-sized tractor factory: the site consists of buildings that have been partly occupied spontaneously to create a cultural center where the brick walls of the old warehouses have been used as support for murals, in partly by production activities and partly nature has reclaimed the empty spaces and generated a third landscape. The project, developed with students from Epoka University's landscape design course, considered the current use as a cultural center and developed a design proposal that supports bottom-up activities proposed by citizens. At the same time, the project preserves the action of nature in the empty areas all around and encourages the development of the third landscape.

Fusha e Aviacionit is a road axis with a very large section whose size is motivated by it originally was an airstrip. On one side the street is lined with a continuous line of high residential buildings, on the other side there is a discontinuous system of empty spaces and buildings (Fig. 6.3).

Fusha and Aviacionit can be interpreted as a linear urban system that connects the green area of the Lana River with the densely built-up city. It interconnects several empty spaces that can be used as parks, where vegetation, with trees that can be planted on the large road surface of Fusha and Aviacionit, can help improve the environmental balance of this densely built-up part of Tirana.

Some case studies are located within two of the city's superblocks where the traces of Tirana's multiform history are most evident and, at the same time, still quite preserved. The first is the super-block bounded by the axes of Rruga Durresit, Rruga Kavajes, and Unaza ring road that ends on Skenderbej Square urban system. The second super-block is bordered by the axes of Rruga e Dibres and Rruga Luigj Gurakuqi-Hoxha Tahsim, the ring of Rruga Bardhyl, and it closes on Skenderbej Square system too. Both neighborhoods are among the oldest in Tirana, close to the physical and urban center of the city and, as mentioned, cut by major road axes. They were also chosen because of their high population density and high vitality.

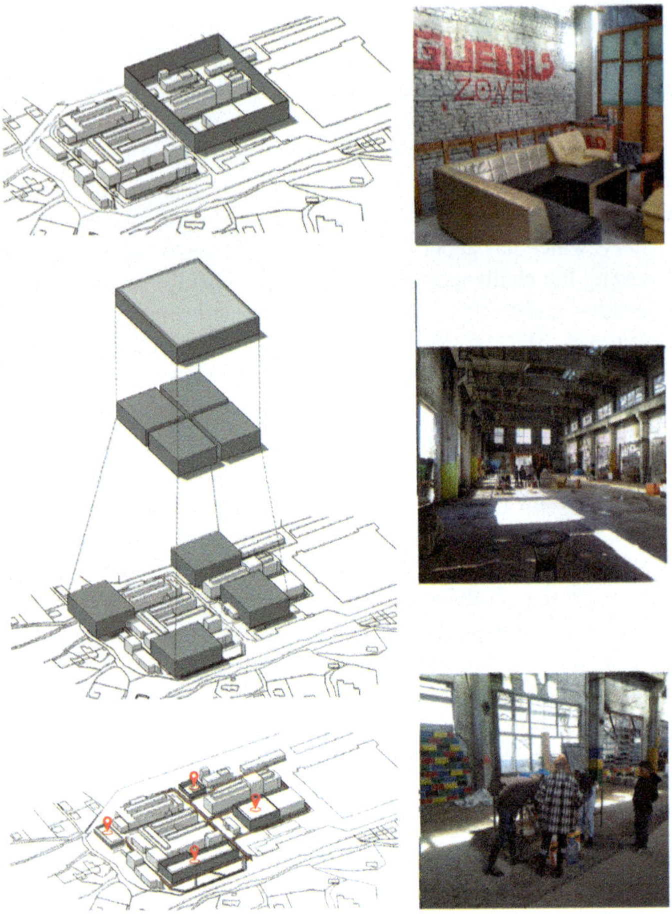

Fig. 6.2 Project concept for the urban revitalization of the abandoned industrial area "Uzina Autotraktora"

This kind of integrated and comparative analysis between the two areas and the urban center was conducted on the topics highlighted above. Specifically, the analysis was conducted on the field, starting with an intuitive and perceptive analysis, such as walks and on-field observations at different times of the day and on different days of the week. There were interviews and meetings with citizens, photo campaigns, and finally questionnaires, composed of questions that were not generic but calibrated on the basis of real local awareness as recorded during the community interaction phases. In parallel, the analysis of data and information received from local and national governments was carried out. Morphological and typological analyses were also carried out following the classic desk-based approach; historical maps and images were compared and correlated with documents, archive and statistical data, and new thematic maps were created from this study. The intent was to map (a) the

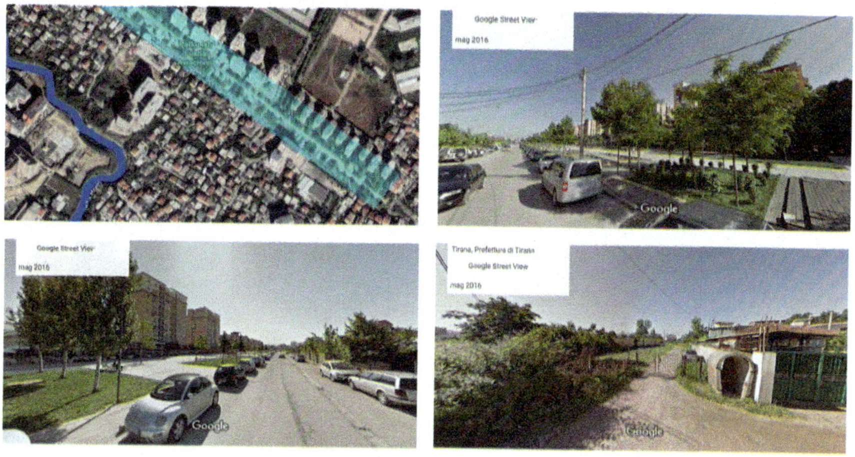

Fig. 6.3 Photographs of the road network of "Fusha e Aviacionit"

residual traces of the fragments of historical urban fabric still present and their state of conservation; (b) the richness and risks of informal adaptations and spontaneous adaptive micro-modifications; (c) the real needs of citizens and their desires; (d) the real consistency of activities and services on ground floors and in open spaces, including sidewalks. Mapping was a useful tool for seeking a balance between real needs and the inalienable values to be safeguarded, in order to recognize, maintain and strengthen the strong urban identities that transpire from these places (Fig. 6.4).

This type of multi-scalar and multi-value mapping can provide useful indicators and indications for any present and future proposal that wants to take due account of the vocations of the places, their physical characteristics, modes of use, accessibility and the activities that take place there. One of the results of the analysis and the corresponding mapping is that we now have a list of wants and needs to be satisfied, through which we can develop project proposals shared with citizens and no longer imposed uncritically. The use of some readily available software has made it possible to consider, for example, data on the actual movements of neighborhood residents, as well as the use (in terms of occupancy) of the space they make daily. The map also reports a similar reading for the city center, namely for the Skenderbej Square system, in order to provide a comparison with the most dynamic and lively part of the city. Additional information concerns the amount of space (especially green space) open to the public, which was found to be severely lacking. The analysis showed that the most recurring type of green space for areas within the superblocks is still the garden, certainly of rural origin and transformed into urban. One of the research projects being conducted on the two superblocks and in the surrounding areas is a design experiment on the area that houses the Radio Tirana building, the radio station of the communist regime. Currently, the building is home to a gypsy family that squats there, while an informal weekly market is organized in the surrounding areas. The revitalization project was developed by co-involving the gispsy family,

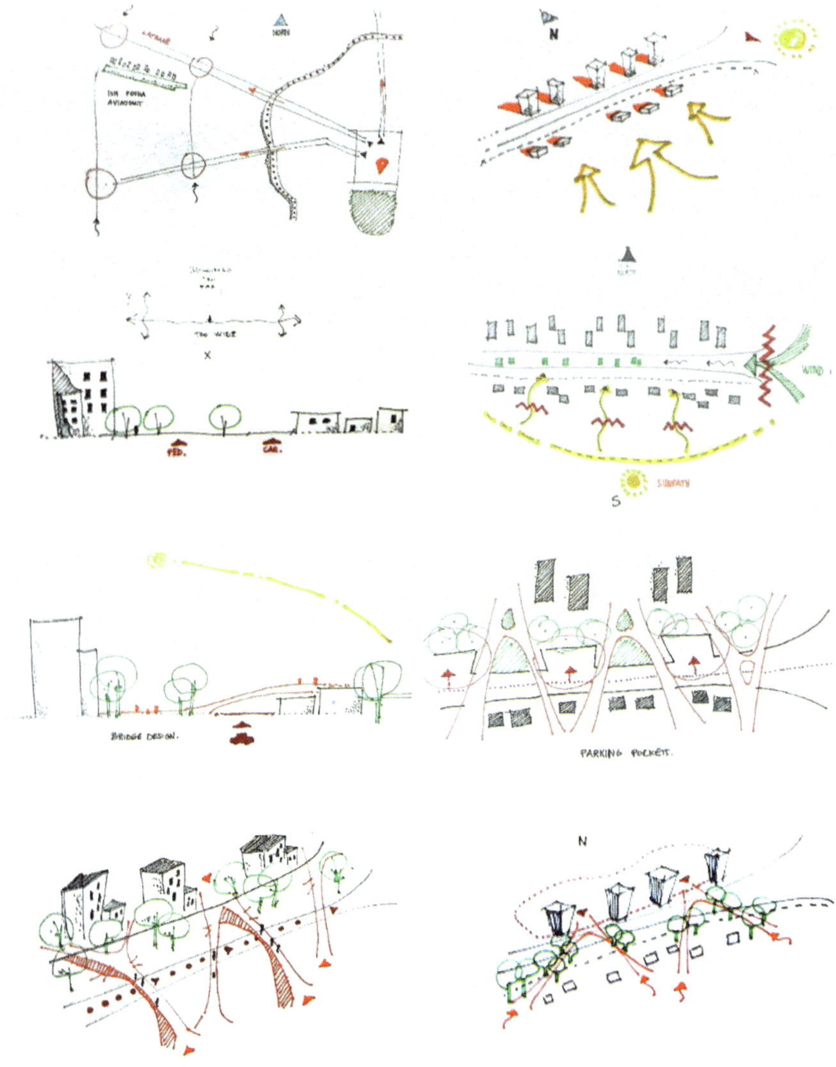

Fig. 6.4 Project concept for the urban revitalization of the road axis of "Fusha e Aviacionit"

who were interviewed and explained what their actual needs and critical issues that they had, how they inhabited the space and what needs they asked to be developed. Functionally, the project resulted in the restoration of the old Radio Tirana building and its adaptation as a residence for the gispsy family. The surrounding area was designed to provide useful equipment for the weekly market. At the scale of the urban design, the area was articulated to provide a partly green public space for the neighborhood and to equip a pedestrian pathway for crossing the inner interlocking areas that make up the neighborhood pattern (Fig. 6.5).

Fig. 6.5 Project concept for the urban revitalization of the abandoned Radio Tirana' and the surrounding area

6.6 Conclusions

Tirana's development in 2030 is officially defined by a set of 13 strategic projects contained within the city's Master Plan, the so-called TR-030, approved and in force since 2018. The official vision of the future city developed by the Plan aims to solve the infrastructural problems of mobility and service networks, environmental and public greenery, public spaces, alternative energy production, regeneration of blighted suburbs, urban neighborhoods on the edge of the consolidated city, peri-urban agriculture, water management and tourism. "Tr-030" consists of three foundational parts that correspond to three stages of Plan drafting, each with a progressive level of depth and definition. The "Metropolitan Fresco," based on ten major themes (biodiversity, polycentrism, diffuse knowledge, mobility, water, geopolitics, tourism, accessibility, agriculture, and energy), releases the overall vision of the plan. The "Atlas," consisting of 13 Strategic Projects divided over the territory, defines the interventions that can trigger the development outlined by the fresco and which will be implemented through a rules document that will represent the third phase of implementation of the plan. Since this is a general plan, the project focuses on an urban scale, rethinking a polycentric, vertical and green city. Not much attention is paid to the scale of neighborhoods and their intangible and informal values. No attention is paid to their intrinsic values and identity attributes, and the land currently occupied by these inner settlements (about 65/70% of the entire city) remains available in the Plan for this strong movement toward verticalization and modernization. Except for some urban redevelopment projects placed in areas between the suburbs and the city, the Plan pursues only a few noteworthy minor conceptual strategies for these inner superblock areas, such as the 24-h use of outdoor school spaces, used and equipped for additional activities to serve the community. This idea is proposed as an implementation of the concept of urban acupuncture, but is not interpreted, for example, in either systemic or participatory terms. Considering also that the Plan was adopted in 2018 and is therefore quite recent, it provides an excellent basis for broader reflections with respect to certain topics and issues that emerged from the research as being of particular importance. The first major issue that emerges in reflections on the current planning instrument on what the plan fails to bring out, is that the priority attention continues to be given, once again in continuity with the formalistic periods in the history of this city, to the "official city" that is, to the city that appears along the boulevards, along the axes and ring roads, from the main squares and public gardens. In this way the plan forgets, or rather guiltily omits, the richness, not only historical, of what we have calculated to represent on average 65/70 percent of the entire city texture: the multifaceted and labyrinthine inner part of the super-blocks. A set of city parts that fill the voids left by the formal city, that gain spaces of uncertainty and sometimes decay, but also of luxury and richness of everyday, very normal, and certainly vibrant activities of the social relations of the natural city. The official city plan offers a broad and strategic vision for developing Tirana as a contemporary and conventional green metropolis. All of the strategies and visions are agreeable but not sufficient.

It is crucial to pay attention to the small and micro scale of the inner city and its discrete spaces, implementing the official strategies and actions with at least three other strategic actions: (1) the definition of the historic center of the city and its ways of protection and management; (2) the mapping of its drosscape and the definition of a network of reclamation and reactivation through criteria of reuse of the empty space in terms of useful emptiness; (3) the definition of the most appropriate and adequate set of interventions (including and especially through bottom-up practices) to protect and enhance the unique identity of Tirana found in the areas that contribute to define the third landscape of the city.

Tirana-Next is an open pilot research—started in 2019 that aims to provide an updated diagnosis based on the analysis of Albania's capital and main city: Tirana. The research studies how the city was in the past, how it has become today, what its current conditions are and what is happening through the current Master Plan "Tirana 2030" and other macro-projects underway to transform the city into a "modern city." Through articulated investigations and surveys within Tirana's neighborhoods, researchers are still exploring and sharing with residents the multiple experiences of the values of interior spaces and places and the changing relationships of humanity and spatiality around them. A series of pilot cases have been studied and analyzed by considering the multiple dimensions of landscape, the architectural and urban heritage and their urban fabric: physical space (solid landscape), social life in the streets (soft landscape), human interactions (human landscape), virtual spaces beyond the physical ones (digital landscape), and the framework of processes (landscapes of cycles and processes). By investigating and analyzing the links between these elements, it is expected, as has been demonstrated so far by the initial results achieved, that this research work can provide a number of recommendations and practical suggestions for a more appropriate systemic regeneration in favor of the real quality of life in Tirana's neighborhoods, even "beyond" the current ongoing formal urban process.

Acknowledgements This chapter presents the research achievements developed jointly by the authors. However, Fabio Naselli's contribution primarily focused on parts 6.2 and 6.3, while Mariateresa Giammetti's contribution specifically concentrated on parts 6.4 and 6.5. Both authors drafted the introduction and conclusions.

References

Aliaj, B., & Lulo, K. (2003). The City of Tirana, history of architecture and urban development. In B. Aliaj, K. Lulo, & M. Genc (Eds.), *Tirana the challenge of urban development*. SLOALBA. Available in: https://www.instat.gov.al

Bulleri, A. (2018). Back to the future. In *Architecture and urban planning for an (extra)ordinary metropolis*. OFL-Oil Forest League.

Clement, G. (2005). In F. De Pieri (Ed.), *Manifesto del Terzo paesaggio*. Quodlibet

Mali, F. (Ed.). (2018). HAPESIRA/ZEROSPACE, Albanian Pavillon 2018. In: *16th International Exhibition of Architecture: La Biennale di Venezia*. RSH Ministria e Kultures.

Naselli, F. (2015). Tourism as a resource in managing the valorisation of the urban and territorial heritage within the Mediterranean basin. MOTRIS, a proposal of integrated relational tourism. *Procedia: Social and Behavioral Sciences, 225*.

Naselli, F., & Trapani, F. (2020). "Ephemeral regeneration" for the marginal urban spaces/places in Enna. In *Proceedings of the XXII SIU national conference "Italian urban planning in front of agenda 2030. Bringing territories and communities to the road to sustainability and resilience"* Matera-Bari, 5–6–7 June 2019, Roma-Milano, Italy: Planum Publisher. ISBN 9788899237219.

Pashako, F. (2016). The legacy of informality in Albanian landscape. In D. Pastore (Ed.), *EVOKED, architectural diptychs*. Giuseppe Laterza srl.

Chapter 7
The Relationship Between Heritage Buildings and Their Urban Unit—Reinterpreting the Role of the Urban Context in Historical Habitats Facing Rapid Transformations, Case of Tirana

Marsela Plyku Demaj, Skender Luarasi, and Blerim Nika

Abstract Our settlements consist of and are formed with urban units. The spatial and formal configuration of these urban units embody the different ways of living in the city as well its transformation in history. Heritage buildings, as permanent features within the urban layout, are endowed with an organizational role in the configuration of their *urban unit* in terms of the formal and spatial relationships that they establish with other buildings within their urban unit. This paper explores these relationships and how they partake in the configuration and sustainability of the urban unit. The paper focuses on the morphological changes that have occurred in the urban unit of the heritage buildings in Tirana, in a time frame of about 60 years. Tirana is the most evident case of rapid transformations caused by both urban sprawl and urban densification. As such the features of many urban units have experienced rapid changes and transformations, often disregarding the context in which these changes are introduced and the existing spatial relationships within them. By using a morphological analysis based on historical maps and data and a comparative reading using interpretative morphological diagrams and models, the paper highlights the main features of the selected heritage buildings. The paper registers the intrinsic formal and spatial connections and relationships that the heritage buildings have established with their Urban Unit(s) in different historical periods and contexts. The paper evaluates the impact of the urban transformation on the heritage buildings: how it sustains, enhances or empowers (the historical value of) heritage buildings, or, on

M. P. Demaj (✉) · S. Luarasi · B. Nika
Polis University, Tirana, Albania
e-mail: marsela_demaj@universitetipolis.edu.al

S. Luarasi
e-mail: skender_luarasi@universitetipolis.edu.al

B. Nika
e-mail: blerim_nika@universitetipolis.edu.al

F. Naselli et al. (eds.), *Self-Sufficient Cities and Sustainable Habitats*, The Urban Book Series, https://doi.org/10.1007/978-3-032-10337-6_7

the contrary, how it weakens, undermines or erases them. The paper also proposes the development of heritage impact assessments prior to any interventions in the context of heritage buildings. Introducing this requirement in policy making and planning enable heritage driven sustainable urban development. The methodology used was introduced and tested in the framework of a workshop with 2nd year architecture students and part of the results produced (models) will be also presented in this paper.

Keywords Historical habitat · Heritage buildings · Heritage impact assessment · Urban morphology · Urban unit

7.1 Introduction

The establishment of today's city of Tirana dates to 1614. The construction of the Old Mosque[1] symbolically marks the founding of Tirana as an urban settlement. The first urban units in the plain of Tirana contained about 15 houses a mosque, a bakery, a hammam and an inn; these constituted the urban core of this settlement.[2] Since its proclamation as the capital of Albania in 1920, it expanded and underwent various urban transformations, during the interwar period, during the socialist period (from 1944 to 1991), and after the 90s. Referring to the city during socialist period, Manahasa and Manahasa (2022), note that *"we cannot say that the city was a socialist city in the purest form, but rather an agglomeration of different urban layers which in many cases intervene, juxtapose or "annihilate" each other for the sake of this political agenda or urgent spatial need dominance"*. However, they further note that, at the end of the socialist period, even though Tirana could be defined as a hybrid of different urban layers, it was still characterized by a quasi-balanced morphology, which changed dramatically after the 90s. Tirana experienced an unprecedented urban explosion in the last 30 years, which took place both as an urban sprawl outside the city and as an urban densification of the existing city. As part of the existing city the urban units have experienced rapid changes and transformations, which disregard the former urban context.

In the last 30 years, research has focused on the analysis of the impacts of the urban development and transformations of Tirana. Dhamo et al (2021) note that, especially after the 90s, urban reality was changing much faster than the urban master plans could help anticipate such a reality. They also argue that drawing up masterplans was no longer even feasible and not even necessary. Marjanaku et al. (2024) note that that the urban transformations have had a great impact on the diminishing of the common public spaces that used to exist within the urban units and neighbourhoods, because of building developments and expansions. In analysing the current situation of Tirana, Luarasi introduces the concept of separate *islands* and *archipelagos* referring to new gated communities and high-rise interventions to today's urban units of Tirana. He

[1] Sulejman Pasha Mosque which today doesn't exist any more.

[2] Frashëri (2004).

Fig. 7.1 Location of selected historical buildings within their urban units in Tirana urban layout

further compares the existing urban layout to a 'sea' between the islands which consists of an 'unbridled urban growth, both as extension beyond the city, and as an erasure of the existing historical fabric from within (Luarasi, 2021)'. Below is a morphological study of the impact of the urban transformation of the urban units of 10 (ten) selected heritage buildings in Tirana, in a time frame of about 60 years (Fig. 7.1).

7.2 Scientific Background and Research Problem

In analysing the morphology of urban units, it is essential to consider a chronology of multiple perspectives that have focused on the complex interplay between physical form, social dynamics and human experience within the city. Lynch (1960) focuses on how people perceive and interact in the city by distinguishing five urban elements (paths, nodes, edges, districts and landmarks), which constitute the city's identity. Jacobs (1993) emphasizes the importance of diverse mixed-use neighbourhoods and pedestrian activity in creating a liveable and meaningful urban unit. Kostof (1991) explores how various spatial patterns and configurations define urban units in different cultural and historical contexts by following an approach based on a historical chronology. Meanwhile, Araújo de Oliveira (2022) follows a more systematic analysis of urban morphology, by examining the physical elements like street networks, blocks and buildings, and spatial organization of urban environments.

His approach emphasizes the analysis of urban form at different scales, from the macro-scale of the city to the micro-scale of individual buildings and streets.

To define space architecturally meant "to determine boundaries". Space has rarely been discussed by architects before the beginning of the twentieth century. (Tschumi, 1975) The boundaries of urban units are visually perceived in terms of formal and spatial relationships, but they are also often determined by the legislation of preservation practices. In the Albanian legislation on cultural heritage the preservation of the 'surroundings' of designated heritage buildings[3] has generally been recognized as a necessary element of the conservation strategy. Generally, a specific protected or buffer zone is drawn around a single historical building. However, due to inflexibility of this normative boundary regarding the urban developments in and adjacent to the buffer areas (no new development is allowed within buffer areas), the protected area is often limited to the same historical building yard or is not designated at all.

The main hypothesis that led to this research is that the premodern and modern interventions of the early twentieth century into the existing organic urban plots were somehow sensitive to the existing context and fit formally with the existing buildings and urban fabric. The opposite seems to happen today when the same buildings (now designated as heritage buildings) seem in many cases to be detached from their urban unit, while the new buildings introduced into the urban units are indifferent to the existing urban historical context. In fact, heritage buildings (being designated as such and separated from their urban unit by a border line drawn in a zoning map) are generally considered as permanent features within their urban units. On the other hand, as Luarasi (2021) points out, considering the dynamic uneven development of the urban units, historical buildings may now be seen like different kinds of *islands* within *archipelagos* that hardly communicate with the newly introduced surrounding.

What is the impact that the transformations of the urban context impose on historical building and vice versa?

In evaluating this impact, we first need to understand the role of heritage buildings in the current urban units, as well as to understand the role that the urban context has in either enhancing the historical values or weakening or erasing them. Distinguishing and understanding an historical habitat, its urban units and its transformations dynamically in history is the methodological basis in this study.

The research is significant not only in the local but also international context, as it contributes in understanding the relationship between a historical building and its urban unit. The research uses archival documentation of historical buildings and the city of Tirana. It also relies on current surveys and documentations through digital mapping techniques, as well as on more 'traditional reading techniques' such as drawing and models.

[3] In the form of buffer zones surrounding the designated building.

7.3 Methodology

The research was introduced and tested in the framework of a workshop with 2nd year architecture students.

From a methodological point of view, the studio exercise proceeds with a two-step approach—(1) analysis and (2) comparison. The exercise encourages the students to look beyond their individual 'plot'/'building'. The method used for defining the urban unit, was inspired from the didactic approach of Angelil and Hebel (2008, p. 233), in which the site-plan is developed as a series of investigative actions.

The analysis of urban units was carried out through the *tracing* and *layering* of the selected urban units in terms of their historical transformation. Then this tracing and layering was evaluated in terms of the changes—continuities or breaks—of the historical relations between heritage buildings and their immediate urban contexts.

Instruction was provided to the students in order to outline a sequence of operations for constructing a sitemap, according to the '*spatial qualities*' of the sites. Participants were divided in groups of 3–4 students and each group was assigned one of the historical buildings selected as case studies and four distinct layers as below were proposed to be developed:

Layer 1—draw the basic structure of the site and buildings using coloured lines.

Layer 2—indicate the most important space-defining architectural elements in your system and record them with overlays in black.

Layer 3—indicate the main functions, as well as secondary functions in your spatial system with coloured fields (served and serving spaces).

Layer 4—consider the impact of light and shadow in your sequence of spaces, programs and uses.

Maps related to four historical moment—those of 1921, 1937, 1958, 2020—were provided to each group (see Fig. 7.2). By using the tracing and layering method, the students identified the historical building(s) in each of the historical maps and speculated on the shape and configuration of their appertaining urban units.

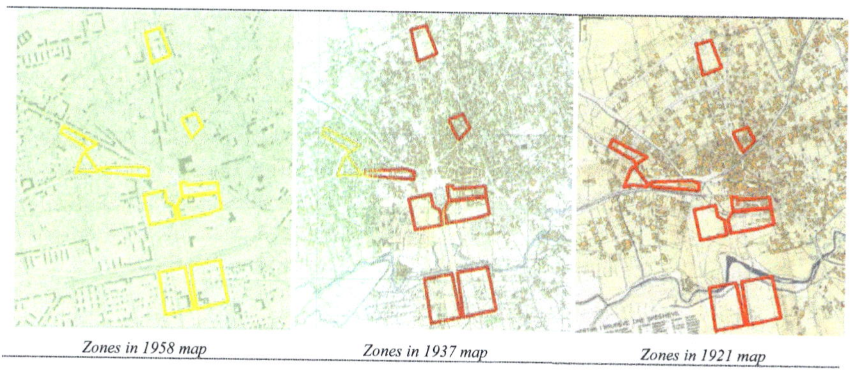

Fig. 7.2 Identification of case study zones in three historical periods

The urban units were analysed in terms of permeability and opacity, served and serving spaces, planimetric and altimetric organization, solid and void, as well as in terms of their visual relationship with the city, namely, how these urban units are perceived from the city, and the latter is perceived from within the units. Then the selected historical building and their connection with the urban units undergoes the same modes of analysis.

The historical maps serve to understand the urban unit as a layered object or palimpsest and its intrinsic relationships in history, while site visits and cognitive mappings of the selected zones serve to project contemporary mental maps onto the palimpsest, thus subjectively positioning oneself in relation to the urban unit. This method of *Serial Visioning,* used also by MUD-Lab (2020), is a visual approach to urban design first introduced by Gordon Cullen in his 1961 classic Townscape. For Cullen, urban design is the 'art of relationships' between the various physical elements of the urban landscape. It is the art of taking 'all the elements that go to create the built environment: buildings, trees, nature, water, traffic, advertisements and so on and to weave them in such a way that drama is released'. It is a presentation from the moving person's perspectives rather than from the bird's eye view. This method of visual representation through photographs and sketches was proposed as a tool for evaluating contemporary experience of moving around a space and understanding it. In the context of hands-on workshop, the students also attempted to create a bridge between history and design by modelling the layered readings of the analysis of the urban unit in two or more historical periods.

7.4 Study Area

7.4.1 Selected Case Studies

Ten historical buildings of the 20s and 30s, which were houses and administrative buildings, were selected as case studies. The selection criteria had to do with the fact that at the moment that they were constructed, they were either the first building in an empty plot (no.1, no.2, no.10), or they were inserted into an existing built context aiming to both reconfigure and fit with Tirana's so-called 'organic' urban fabric. It was in these first years of the twentieth century, and after the proclamation of Tirana as capital, that significant urban interventions were introduced in the existing historical habitat. Considering that the main stock of the protected historical buildings today in Tirana consists of early twentieth century-built heritage, the focus of the research was this period. The selected historical buildings are part of the National List of

Fig. 7.3 Selected historical buildings

Cultural Monuments since based on their architectural and historical values, they were designated as 2nd Category Monuments[4] at 2007 (Fig. 7.3).

[4] Based on the Albanian legislation for cultural heritage, the 2nd Category Monuments are generally protected as complementary to the already protected urban ensembles or urban units as well as for (but not limited to) the values presented in their exterior, volumes, composition of the plans, facades, decorations, etc.

7.5 Results

The result of the studio confirmed the initial hypothesis that in most cases, the uneven and insensitive new urban developments have contributed to the transformation or disfiguration of the urban units. Based on presentations of their findings at the end of Task 1, the students were able to explain their analysis in each of the layers and in each of the historical periods.

Even though the selected historical buildings are conserved as permanent features within the urban units, the urban units have undergone deep transformations. The latter are visible in all the layers extracted during analysis. While starting from a denser network of lines depicting either streets or building lines (Fig. 7.4), the current situation results in less traces and layers, which evinces erasure of the former fabric and more impenetrable large buildings.

Analysis of the layers depicting especially in common and private spaces showed a shift in building scale and a diminishing of common space in favour of private properties (Fig. 7.5). Shadow analysis displays the drastic changes and shifts in the relationships among the buildings comprising the urban unit (Fig. 7.6). In many cases, the new buildings appear as separate elements within what used to be a distinct urban unit, ignoring the relationship with their surroundings and adjacent structures. Such disfiguration of the urban unit has a negative impact on the historical buildings (even if they are not destroyed), insofar as it destroys the urban space in and from which we can experience these historical buildings.

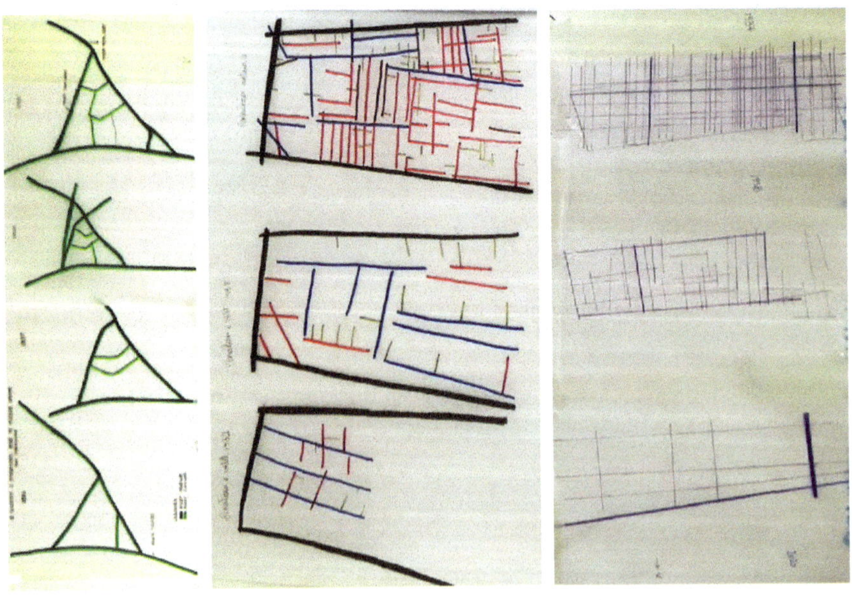

Fig. 7.4 Examples from the layer analysis undertaken in Task 1—Layer 1

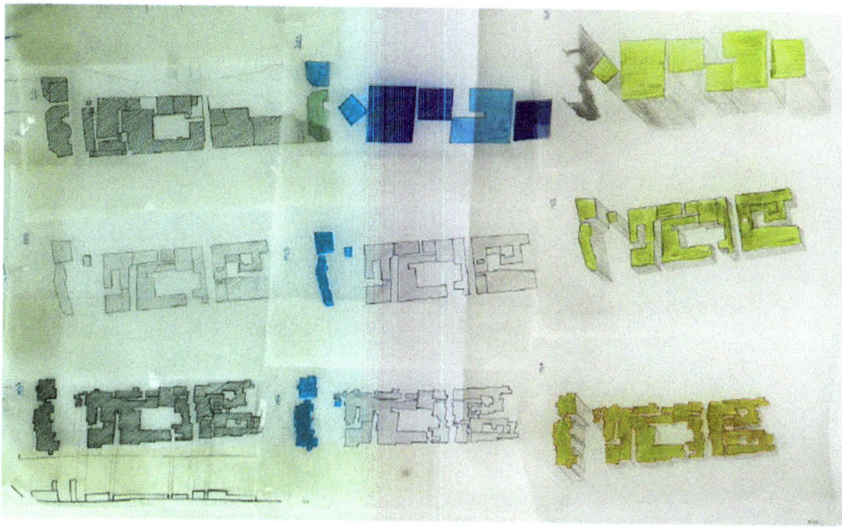

Fig. 7.5 Examples from the layer analysis undertaken in Task 1—Layer 2–3-4

The second task focused on the comparative analysis of the urban unit. Based on the selection of two historical periods that best represent the changes in the relationship between historical buildings and their urban unit, students modelled the results derived from the layer analysis performed in Task 1 (Figs. 7.6, 7.7, 7.8, 7.9, 7.10 and 7.11).

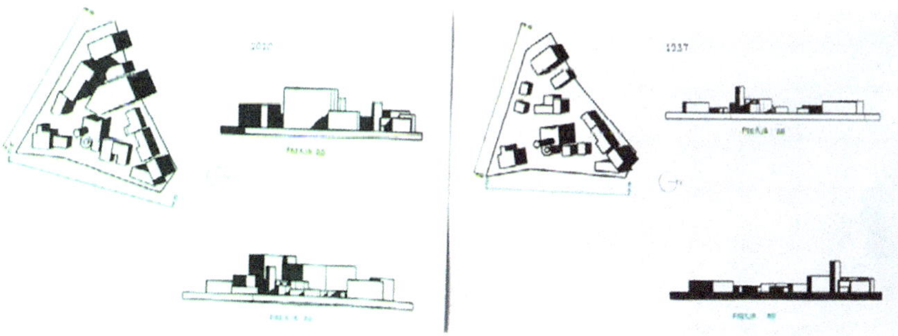

Fig. 7.6 Examples from the layer analysis undertaken during Task 1—Layer 4

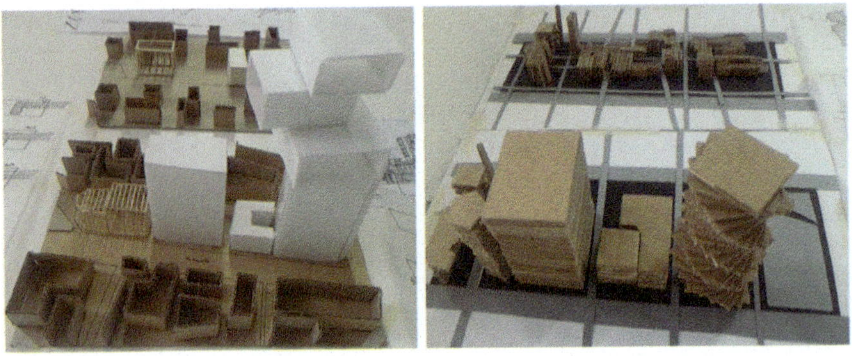

Fig. 7.7 (right)—Final model Zone 6—Former Writers League, (left) Final model, Zone 4—Municipality of Tirana

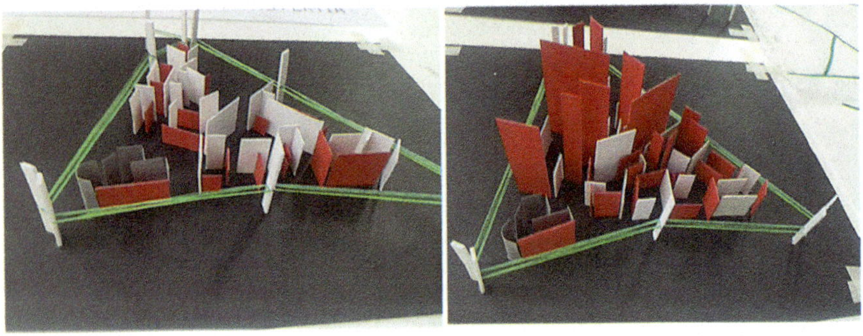

Fig. 7.8 Final model Zone 7—Former Pastërtia Building

Fig. 7.9 Final model Zone 5—Former Ministry of Environment and Tourism

Fig. 7.10 Final model Zone 8—Ministry of Education

Fig. 7.11 View from the presentation of the results

7.6 Conclusions

The study evaluates the impact of urban transformation on the heritage buildings, in the context of Tirana. The study shows that the historical buildings are heavily impacted by the new, detached developments that disfigure the urban unit and the historical habitat in which the historical buildings are situated. These new developments detach the historical buildings from the urban units by disfiguring their relationship. Based on the definition of Augé (2008, p. 63) who states that *'a space which cannot be defined as relational, historical or concerned with identity, will be a non-place'*, it can be concluded that the current approaches of the urban developments in the city of Tirana are unfortunately nurturing the creation of *non-places*. In this context, this study also supports the conclusions drawn by Luarasi (2021, p. 92) who states that what is at stake today *'precisely the re-working, re-acquirement or retrieval of the very concept of the whole, both beyond a nineteenth century organicism and postmodern fragmentary dispersions'*. This study suggests that the search for solutions needs to focus on those urban units that still conserve historical layers. In doing so, such urban units need to be re-evaluated by using a holistic and integrated approach that takes into consideration the context of the historical buildings. The buffer zones of historical buildings should not prevent but carefully manage change in the urban units by using a value-based approach.

Acknowledgements The authors are acknowledging the contribution of Arch. Joli Mitrojorgji who provided assistance and valuable insights to the students during their presentations at the Workshop.

References

Angelil, M., & Hebel, D. (2008). *Deviations—designing architecture—a manual*. Birkhauser Verlag AG.

Araújo de Oliveira, V. M. (2022). The elements of urban form. In U*rban morphology. The urban book series*. Springer. https://doi.org/10.1007/978-3-030-92454-6_2

Augé, M. (2008). *Non-places—an introduction to supermodernity*. Verso.

Dhamo, S., Thomai, G., & Aliaj, B. (2021). *Tirana Qyteti i Munguar.* Pegi

Frashëri, K. (2004). *Historia e Tiranës*. Toena.

Jacobs, J. (1993). *The death and life of great American cities*. Vintage Books.

Kostof, S. (1991). *The city shaped: urban patterns and meanings through history*. Little, Brown and Company.

Luarasi, S. (2021). *Albanian archipelagos, the city of a non-normativity foretold*. FORUM A+P. https://doi.org/10.37199/f40002211

Lynch, K. (1960). *The image of the city*. MIT Press.

Manahasa, E., & Manahasa, O. (2022). Nostalgia for the lost built environment of a socialist city: An empirical study in post-socialist Tirana. *Habitat International, 119*. https://doi.org/10.1016/j.habitatint.2021.102493

Manchester Urban Design LAB. (2020). MUD-Lab Toolkit: Serial Vision' accessible at www.seed.manchester.ac.uk/mudlab

Marjanaku, H., Plyku Demaj, M., & Kumaraku, L. (2024). Exploring the balance between common and private spaces. A case study from Tirana. In *Proceedings: Issues of housing, planning, and resilient development of the territory towards euro-mediterranean perspectives*. https://doi.org/10.37199/c41000109

Tschumi, B. (1975). *The architectural paradox, studio international; 1975; revised in architecture and disjunsction (1994)*. MIT Press.

Vokshi, A. (2013). Tracce dell'Architettura italiana in Albania 1925–1939. Ph.D. Thesis, accessible at https://flore.unifi.it/handle/2158/799873. Technical Archive of the National Institute for Cultural Monuments (Monuments' Files).

Chapter 8
Laic and Religious Areas in the Small Town of Vigonza (Italy). Examples of Urban Mending Projects

Enrico Pietrogrande, Alessandro Dalla Caneva, and Massimo Mucci

Abstract The historic centre of Vigonza, a small town in the Venetian area, has the peculiarity of presenting the town hall and the church located in separated and distant places, while generally these landmarks for public life stay in the same area of the urban tissue. Vigonza is one of the subjects recently investigated by our students at the Architectural and Urban Composition 2 course taught on the master's degree in Building Engineering—Architecture at the Department of Civil, Environmental and Architectural Engineering of the University of Padua. The centres of the town of Vigonza were thought by our students as an opportunity to configure the unity of these symbolic and representative places. The analysis and design methodology make use of the tools developed on typology and urban morphology studies by the Italian rationalist school aimed to search a meditated dialectic between the new and the old in the urban project. The students' proposals are basically focused on the new route which, like the string of a bow, connects the two important poles of the town, the administrative and the religious one. The intention was to reinforce their identity. The projects also involved a large green area located at the south of the path connecting the town hall and the church through a straight line. The research experience has shown that the contemporary urban design can insert new elements in the preexistent context while respecting the urban identity.

Keywords Identity · Memories · Urban form · Urban setting · Urban place

E. Pietrogrande (✉) · A. D. Caneva · M. Mucci
Department of Civil, Environmental and Architectural Engineering, University of Padua, Padua, Italy
e-mail: enrico.pietrogrande@unipd.it

A. D. Caneva
e-mail: alessandro.dallacaneva@unipd.it

M. Mucci
e-mail: massimo.mucci@unipd.it

F. Naselli et al. (eds.), *Self-Sufficient Cities and Sustainable Habitats*, The Urban Book Series, https://doi.org/10.1007/978-3-032-10337-6_8

8.1 Introduction

While the fundamental landmarks for the community in Italian towns are generally located in the same area of the urban tissue, almost always in the same square, the town hall and the church of Vigonza are contained in the historical centre but peculiarly for Italy these are situated in separate and distant places (Fig. 8.1). Very unusually, in the case of Vigonza the buildings aimed at consolidating the new kingdom of Italy, above all the schools and the town hall, were not located close to the Church of Santa Margherita at the end of Via Roma, but along what is now Via Cavour (Benetti, 2000). This choice was probably made to mark the secular origin of these institutions. The area around the town hall was further developed in the late 1930s (Pietrogrande, 2011) when a new square surrounded by new buildings now named Piazza Zanella was developed according to a masterplan proposed by the architect Quirino De Giorgio (De Giorgio, 1940), and this building complex was built between 1937 and 1939 and naturally weakened the religious pole. The almost exclusive use of exposed terracotta and the abstract design of the main façades give the place a metaphysical character. A secondary consequence was that the axis of Via Roma, which ends in the historic Church of Santa Margherita, was deprived of a definite identity.

The secular administrative pole (Fig. 8.2) is dominated by the town hall (Fig. 8.2) and the theatre (Fig. 8.3). De Giorgio personally designed the theatre, the covered market (Fig. 8.3) and other buildings thought to host a few peasant families relocated from unhealthy housing (Rigamo, 1938). The architecture of the theatre, almost completely realized with exposed bricks, has a convex façade which, combined with

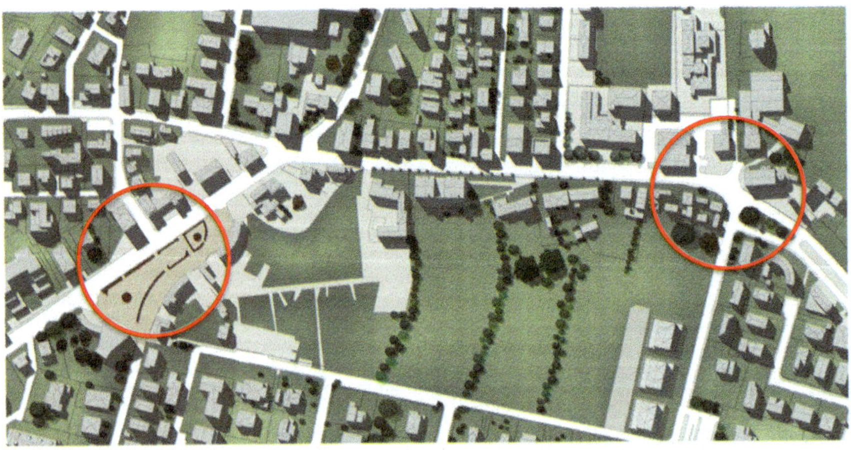

Fig. 8.1 Planivolumetric representation of the centre of the town of Vigonza. The study area is included between the square named Piazza Zanella hosting the secular administrative pole on the left in the image, and the religious pole composed of the Church of Santa Margherita on the right. Piazza Zanella can be recognized from its sinusoidal shape and the church because it is located at the end of Via Roma

the concave profile of the adjacent covered market, gives rise to a sinusoidal shape. In the central part of the façade there is a tall entrance portico, framed by the compact building's masonry facing and divided into six parts by five columns, also finished in exposed terracotta. The joints between the brick courses in the columns are almost nullified by the peculiar shape of the single pieces, which have the bases tapered so as to accommodate the cement in the inner part only: the result is the smooth surface of the cylinders, and their clear reaction to light versus shadow in depth.

During the Fascist period, the government's efforts to improve the living conditions of the population in the countryside intensified. Mussolini opposed immigration from the countryside to the towns and cities, a policy chosen to avoid overpopulation of the cities which would have made the control of the population and the maintenance of social order more difficult. In 1934, he thus synthesized his policy during the 15th anniversary of the foundation of the 'Fascio di Combattimento' (a Fascist Party organization) in the following way: 'Within a few decades all Italians living in the country must have large, healthy habitations where the farming generations can live and persist over the centuries' (Pennacchi, 2011).

The houses included twelve dwellings and stores on the ground floor. Their volumes, also arranged in a sinusoidal pattern, were arranged on the north side

Fig. 8.2 (top right) Laic area in the town of Vigonza. (left) Map including the square now named Piazza Zanella, with the town hall and the two sinusoids constituted by the theatre, the covered market and the houses for peasants. (bottom right) Views of the peasants houses and the town hall

Fig. 8.3 (left) and (right) View of the main façade of the theatre with the arcades of the covered market to the side and the interior of the covered market

of the square. All buildings, except the central one, have the façades finished with exposed bricks and the roofs descending to the rear so that the volumes appear taller. The roofs of the porches are finished in tiles like those of the houses, confirming the uniformity of the overall colour scheme. The façades are characterized by the repetition of a basic module consisting of the rectangles of windows cut out on the second floor on the masonry and the corresponding portion below. The front of the farthest building from the theatre is enriched by a relief sculpture centred on the themes of rural work and motherhood, proper themes to the rhetoric of fascism.

The way of composing architecture by juxtaposition of elementary volumes, practiced by De Giorgio markedly in these years, stands out in the photographs. The changes made to the buildings of the secular area over time are circumstantial and, thanks to the period photographic documentation, clearly recognizable. On the whole, the buildings in the square have been preserved free from traumatic alterations, and still day they express the tension and organicity of the original project.

The west of the space of the religious pole at the end of Via Roma is delimited by the church as well as other buildings which are important references for the community (Fig. 8.4).

The church, designed by the architect Jacopo Sacchetti, had been built between 1834 and 1837. It was the third building with such a destination after two previous ones insisted on the same site (Nesoti, 2007). The façade was rebuilt in brick and stone following the damage produced by aerial bombing on the 20th of March 1945, in an advanced position so as to enlarge its covered surface.

On the north side, near the church, is the volume of the old building used for parish activities presenting a fascinating historical character. Now abandoned, it dates back to no later than the seventeenth century. The most important building in the local religious tradition stands on the area behind the church. It is the cloister of the former convent of Santa Margherita, which has roots dating back to the twelfth century. The original church and convent can be seen in documents from the years 1138 and 1155 (Frassine, 2019).

Fig. 8.4 View of the religious pole of the town of Vigonza. The Church of Santa Margherita with its bell tower is at the centre of the figure, on the left is the decommissioned charitable institution of San Sebastiano and on the right is the convent whose layout dates back to the twelfth century

Project proposals focus on the new path, already partly implemented, which like the string of an arch connects the two poles of the town, the secular and the religious. In particular, the proposals aim to create a relationship between the intermediate part of the path and the southern front of Via Roma to help consolidate the identity of the latter. In fact, while Via Roma is one of the oldest routes in the urban area, and while effective in establishing an operational link between the various parts of the town, it now lacks a unified character. This results from the mixture of heterogeneous recent volumes and significant historical buildings of both minor and of monumental value—among the latter the presence of a late sixteenth-century palace attributed to Vincenzo Scamozzi should be noted. The buildings house commercial activities or crafts on the ground floor and housing on the upper floors. The built profile lacks order and coordination so Via Roma appears to have lost its original relevance within the urban fabric. This needs to be remedied in view of a primacy that can be restored. A large green area composed of a park available to the community lies to the south of the straight line that connects the town hall and the church. Developing a relationship between the southern side of Via Roma and this vast green area is another possibility offered by the site.

Although not documented in this article, other project themes in Vigonza have been the redevelopment of the access road from the city of Padua and the improvement of the area of the old mill and castle in the nearby hamlet of Peraga (Benetti, 1975). These are different components of issues traceable to the broader theme of redevelopment implemented in a specific context through both the transformation of brownfield sites and through the improvement of qualifying aspects. Generally, the belief that the redevelopment of the environment must progressively replace the consumption of virgin land characterizes the teaching of the authors in addition to the principle that the study of place is the indispensable initial tool in developing the project.

8.2 Methodology

The project proposals presented in this article consider the city as the reference point for rethinking the urban space. They are experiences that are not isolated but embedded in a wider historical context whose character needs to be interpreted. Urban analysis is the method adopted to understand the compositional principles through which new project experiences rooted in place are constructed. This attitude agrees with the typological and morphological studies of the city made by the Venetian school with particular reference to both the writings of Aldo Rossi and the more recent experiences of so-called neo-rationalism. According to this trend, the enhancement of space and the values encapsulated in it are a prerequisite for a society that is not only useful but also beautiful.

This school considers the city as a spatial phenomenon. In this regard, the words of Uwe Schroder are meaningful: 'Growing and gaining ground in the field of architecture—a discipline that is generally occupied with building on and in spaces—is the conviction that a more substantial consideration of the spatial design conception of the buildings and the city may not only reflect on the form necessary but also lead to an approach capable of solving various current social questions' (Schröder, 2017).

The fundamental approach to the subject of the historic city in the projects developed by the students is based on understanding the space through its typological and morphological forms and devices (Fig. 8.5). Even though it is the vocation of architecture, this attention to space is often downplayed by a tendency to produce self-referential objects incapable of contributing to a more civilized urban form. It is an approach that is a continuation of modernist achievements, faithful to the patterns introduced in the twentieth century by which architecture is reduced to a spectacular object placed in a space consequently lacking quality (Caja, 2017). As an alternative to a certain reformist line of argument founded on the idea that urban redevelopment is the substitution of parts of the existing city, the theoretical position referred to in this paper examines the compact historical fabric to generate the principles using which the urban design is developed.

This idea is in contrast to the concept of sustainability understood only as technological performance aimed at reducing environmental impact. In fact, it pursues a

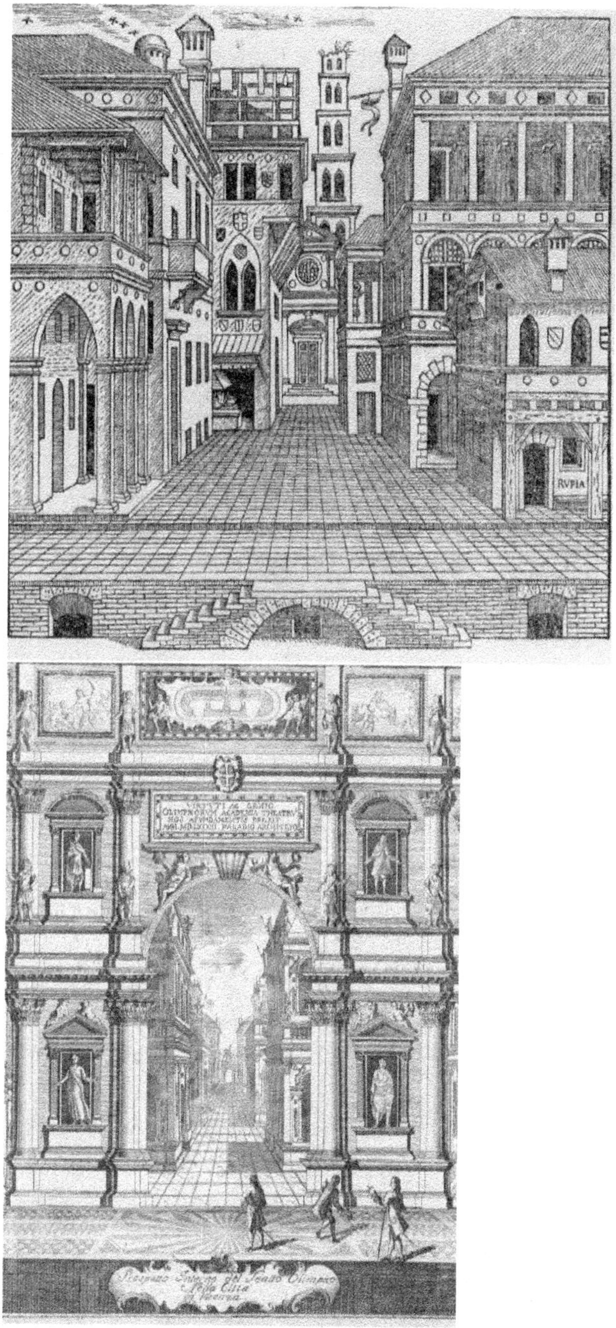

Fig. 8.5 Woodcut illustration by Sebastiano Serlio from the second book of the Trattato di Architettura, mid-sixteenth century (left). Andrea Palladio, Vicenzo Scamozzi, Olympic Theatre in the city of Vicenza, interior elevation, late sixteenth century (right)

social type of sustainability, where the focus is shifted to the need of identity felt by the community. In this sense, it is important to maintain the continuity of the spatial memory of the village, which in the case of Vigonza is to be reconstructed through careful urban mending interventions.

Urban regeneration plans can become a viable tool for building urban form when architecture is seen as a tool that deals with the construction of the quality of space, as a discipline that arranges volumes in space by designing relationships between physical forms. Good urban projects cannot ignore the problem of the quality of urban spaces. It is the forms of urban spaces, the spaces of squares and streets, that ground the idea of the city (Fig. 8.6). As Carlo Moccia said, taking up Uwe Schröder's studies of spatiality, 'urban form has to be investigated in its spatial value, measuring the degree of internal or external quality that connotes the spaces of the city and recognizing the relationships between spaces, of different character and hierarchy, that organically make up the space of the city' (Moccia, 2015).

The tendency to regard historic city spaces as still valid as places suited to satisfy the rhythms of life of contemporary man has deemed the relationship with theories of historic city construction fundamental as a method of intervention. The following words of Antonio Monestiroli are valid in this regard. 'The closed city is defined not only by the presence of the walls, which are certainly the most obvious element of this model, but is defined primarily by the fact that it corresponds to urban places understood exclusively as interiors spaces (the street, the square) in total separateness from natural places. All urban history in the Western world is the history of the construction of the enclosed city: from the ancient city, through the mercantile city, to the Renaissance city, the street and the square are places of presence of the urban elements that themselves represent there and from there are recognized (…). The closed city has developed a formidable system of construction, which is that of the

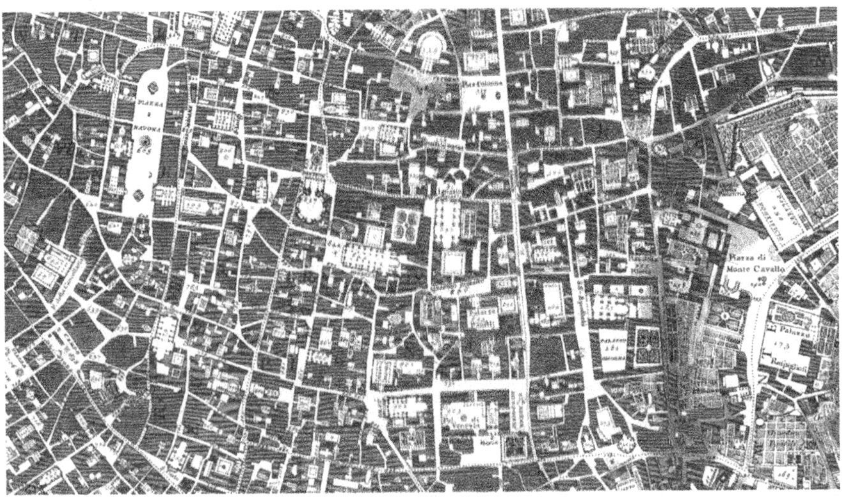

Fig. 8.6 Giambattista Nolli, New map of Rome, 1748

urban block, a portion of land surrounded by streets, which becomes the elementary part of city construction' (Monestiroli, 1997).

In the reported teaching experience, therefore, interest was turned to the forms of construction of urban space, with reference to the courtyard, the portico, the loggia, the gallery, the street and the garden. These devices represent a recurring distinctive character in the form of the historic city, typological forms to look to in order to promote the reconstruction of unresolved areas within the ancient city.

8.3 Project Proposals

Some examples of the teaching experience in the study of the centre of Vigonza are summarized below. They were developed on the Architectural and Urban Composition 2 course at the Department of Civil, Environmental and Architectural Engineering of the University of Padua taught by Professors Enrico Pietrogrande and Alessandro Dalla Caneva.

In general, the projects aim to reconnect the two urban poles in various ways, the secular administrative pole with the religious one by mending the urban tissue and its functional areas. The strategy of rethinking public space by inserting new urban architectures able to provide empty spaces with form is shared by the various project proposals.

Sustainability here is expressed according to principles other than energy efficiency or low environmental impact of materials. These principles are now taken for granted and incorporated into current building practices in accordance with increasingly strict regulations. Here, reference is made to the sustainability inherent in urban regeneration that has at the core of its compositional practices the reconstruction of urban spaces including meaningful architectural quality and, in this case, linked to the local tradition (Musco, 2009). In particular, the reference to urban traditional tissues means building high density by giving priority to the relationship between architecture and pedestrian public spaces, but without abandoning the use of contemporary architectural languages. The goal is not a nostalgic imitation of the past, but a reconstruction of the architectural spatial unity that was typical in pre-industrial cities. In this way, certain assumptions of sustainable urbanism are also realized, such as the reduction of land consumption through more compact and optimized land use.

The students Gabriella Coletta and Andrea Giagnorio suggest continuing the existing path that starts on the side of Zanella square (Fig. 8.7) in the direction of the church of Santa Margherita, that is, crossing the current green area. This tree-lined axis is crossed by a combination of a new building housing the municipal library together with several pre-existing historic mansions, all of which are arranged to form a new pedestrianized square, which is consequently free from vehicular traffic (Figs. 8.8). This square constitutes a third pole distinguished by its cultural character being enhanced by an archive dedicated to architect De Giorgio. It allows the long path between the town hall and the church to be moved to the south and immersed in the urban park. Using double-height arcades and exposed brickwork in the facades gives the new building in the square a public character and echoes the architectural features of the existing urban fabric, specifically referring to Quirino De Giorgio's intervention on Zanella Square. The library gives the square a symmetrical layout thanks to its two projecting wings punctuated by double-height pillars. The square includes the existing historic villas and partially faces the church.

Students Sofia Tarousa and Carlotta Toniato focused on the redevelopment of the area by proposing a new square with a cultural centre facing directly onto Via Roma in an intermediate position between Piazza Zanella and the church. The new building complex is seen from L-shaped when seen from above (Figs. 8.9, 8.10). The arcades and some covered walkways facilitate the passage of people in the direction of the urban park to the south. Consequently, the building complex becomes a connecting point between Via Roma and the park as well as between Piazza Zanella and the

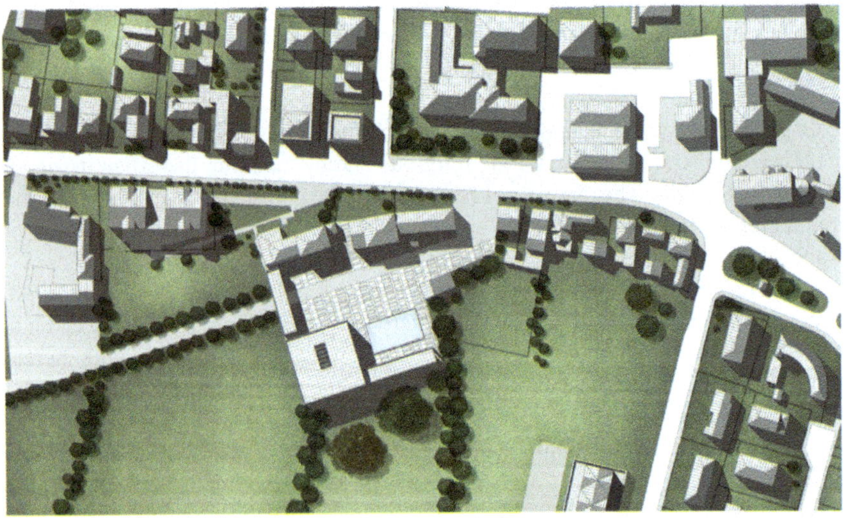

Fig. 8.7 Project planivolumetric plan of the new central part of Vigonza. From the work of the students Gabriella Coletta and Andrea Giagnorio

Fig. 8.8 View of the new square and the library. From the work of Gabriella Coletta and Andrea Giagnorio

church. The project involves the creation of a cultural centre with multifunctional spaces for various activities. The core of the project is a new Creative Hub, with the main goals of stimulating minds and promoting economic development. The Creative Hub contains spaces for co-working, studios, art workshops, rooms for cultural events, a library and a museum dedicated to Quirino de Giorgio. The architectural language consists of elementary volumes and the exposed brickwork of the basement juxtaposed with the white plaster seen above floor level (Fig. 8.10). Lastly, a certain degree of abstraction is combined with the revival of the local building tradition.

A third example of the educational results achieved during the course is the project of students Francesco Mauro and Veronica Merlo (Figs. 8.11, 8.12). In this case, the

Fig. 8.9 Project planivolumetric plan of the new central part of Vigonza according to the proposal made by Sofia Tarousa and Carlotta Toniato

Fig. 8.10 New square proposed by Sofia Tarousa e Carlotta Toniato closed to Via Roma. Plan of the buildings and view of Via Roma

idea was to start from the alignments of the street fronts of the existing buildings between Via Roma and the park. As in the previous projects, the fundamental decision to create a direct connection between the two opposite poles of the town hall and the church is confirmed, but here, the students focus on reconnecting and completing the area of the existing villas with a new cultural centre divided between several new coordinated buildings. Rotations and plan irregularities of the individual buildings are due to the misalignments of the existing fronts of which they are an extension. The volumes proposed are balanced within the well-defined general urban morphology in the intervention area. In this case, the buildings do not define an actual square but rather a web of pathways connecting Via Roma with the urban park.

The didactic experience through the morphological study of the urban area has therefore resulted in a series of projects that contain common characteristics. First and foremost, the desire to counter the volumetric sprawl through the repurposing of simple urban morphologies anchored in context. Besides, the attention to a modern architectural language derived from the pre-war rationalism, yet including geometric

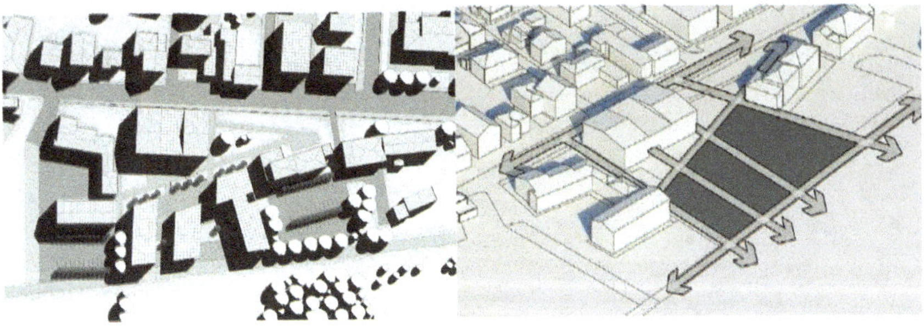

Fig. 8.11 Project planovolumetric plan of the new central part of Vigonza and basic scheme according to the proposal made by Francesco Mauro and Veronica Merlo

Fig. 8.12 General view of the new buildings proposed by Francesco Mauro and Veronica Merlo

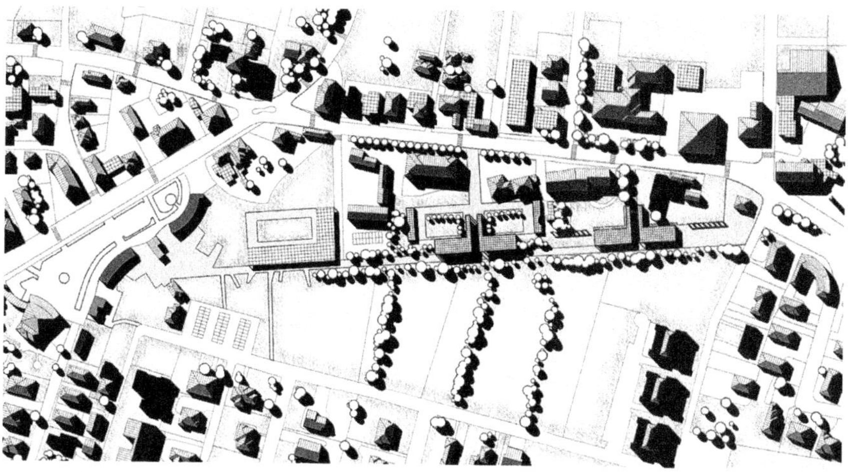

Fig. 8.13 Project planivolumetric plan of the new central part of Vigonza from the work of the students Elisabetta Presa and Sara Simioni

simplicity and material reference to local tradition. Thus, the common approach towards urban redevelopment consists on the research of the best way to connect the project to the context through the study of history and the critical investigation of current functional needs, as an expression of the urban sustainability.

The solution proposed by two other students, Elisabetta Presa and Sara Simioni, consists in the new connection between the two poles marked by three building cores, the covered market, two buildings with a symmetrical arrangement intended for housing and two others similar in form used for public activities (Figs. 8.13, 8.14). The new buildings have the portico as their main compositional theme. Walking along the new axis, one will be accompanied on the one hand by the regular cadence of the pillars, and on the other by a continuous front of trees and the natural landscape. In addition, through three orthogonal paths, the park is easily accessible from Via Roma and it is more integrated into the daily life of Vigonza.

The new buildings recall the architecture of the square designed by Quirino De Giorgio, which presents façades in exposed brick or plastered with white colour, in addition to the arcades that are the common element throughout the project.

Lastly, Jacopo Cassandro and Daniel Trento focused their attention on the unresolved current state of the religious area, which lacks a coherent design, making it the space from which the long new pathway, which implements the relationship with the secular pole, originates and in which it ends. The pathway is characterized by an exaggerated rectilinear course. The new square is designed in front of the church and the patronage of San Sebastiano (Figs. 8.15, 8.16). The latter is the pivotal element of the project, as it defines the conformation of the square. Since this building was originally tripartite, having over the years undergone the demolition of the portion at the west side, the reconstruction of the missing volume is proposed with a wood finishing that characterizes all the volumes added in the project. As a result, the

Fig. 8.14 Elisabetta Presa, Sara Simioni, view of an access to the park coming from Via Roma

patronage of San Sebastiano acquires a dual soul: the more traditional one, maintained unchanged, made of brickwork finished with plaster and roofing tiles, and the reconstructed one, characterized by wooden walls and a sheet metal roof.

A slight modification of the road system, particularly regarding Via Roma, was necessary. The proposal presented by Cassandro and Trento includes new road crossings, the change of some parking lots and the demolition of some buildings along the

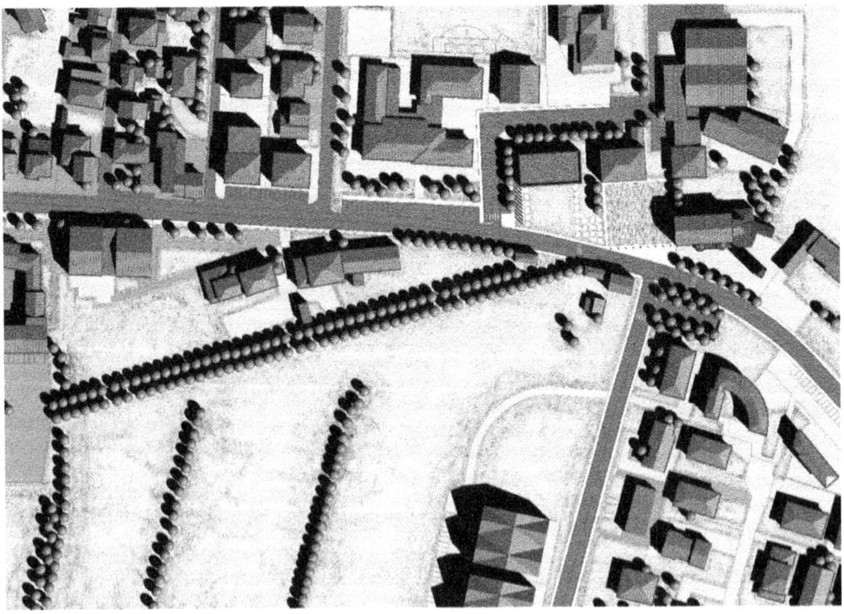

Fig. 8.15 Project planivolumetric plan of the new central part of Vigonza from the work of the students Jacopo Cassandro and Daniel Trento

Fig. 8.16 (top) and (bottom) Jacopo Cassandro and Daniel Trento. Views of the church and the new square

south front of Via Roma, poor quality constructions that were not useful in generating a unified street front.

8.4 Conclusions

The project experiences conducted on the Vigonza city context provide an opportunity to rethink parts of the city by giving them a new role and meaning. It is not a matter of rethinking the area exclusively with new functions, but also with forms that can dialogue with the settlement pattern of the city by building urban spaces in which a community can recognize itself. Architecture is required to enter the context by discretely interpreting the settlement rules in it, the characteristics of the place, and its atmosphere. By constructing spatial sequences by arranging volumes—substituting those demolished—in continuity with the existing ones, respecting their alignment as much on the horizontal plane as on the vertical. There is no need for the form of the planning project to mimic the historic form of the city. Above all, the form needs to interpret the reasons for building the space in which the planning project is to be inserted. This is a possible urban sustainability strategy that leads to increasing the sense of community in the population. In fact, the mending of the urban fabric corresponds to the elimination of urban voids that are experienced as interruptions in the community's spatial memory.

For this reason, the design, which moves on the plane of form, cannot be too stiffened by the norm, since this entails, as history teaches, giving up designing the modern city. Today, urban transformation modes show that they have abandoned this purpose. Plans, laws and regulations are no longer committed to thinking about a regulatory system appropriate to the aesthetic purposes of architecture. Regulations ended up fixing quantities without giving information about the form of places. Architecture has thus been uprooted from its proper role, that of shaping the city through a clear formal idea. In an abstract way, laws and norms were gradually produced that defined ways of doing and building that no longer concerned the physicality of the city.

The projects presented should be interpreted as a valid way for reconstructing urban form by considering architecture as a tool that deals with the quality of space, as a discipline that arranges volumes in space and weaves relationships between physical forms, also in respect to the current urban sustainable principles. The design experiences developed by the students looked to this perspective.

References

Benetti, A. (1975). *Peraga e le Pievi del Graticolato Romano* (Peraga and the Pievi of the Graticolato Romano), Collegio missioni africane, Verona.

Benetti, G. (2000). *Storia del territorio vigontino* (History of the territory of Vigonza), Comune di Vigonza, Padua.

Caja, M. (2017). *Ricostruzione critica come principio urbano e altri scritti* (Critical reconstruction as an urban principle and other essays), Aion Edizioni, Florence.

De Giorgio, Q. (1940). *Tre anni di marcia del fascismo padovano* (Three years of the march of Paduan fascism), Società Cooperativa Tipografica, Padua.

Frassine, M. (2019). *Vigonza. Dal sacello romano all'insediamento medievale* (Vigonza. From Roman sacellum to medieval settlement), Società archeologica, Quingentole (Mn).

Moccia, C. (2015). Come radure nei boschi (Like glades in the woods) in "I due elementi dell'edificazione dello spazio. Scritti scelti", U. Schröder, Aion Edizioni, Florence, pp. 9–13.

Monestiroli, A. (1997). *Temi urbani (Urban themes)* (pp. 10–11). Edizioni Unicopli, Trezzano sul Naviglio (Mi).

Musco, F. (2009). *Rigenerazione urbana e sostenibilità (Urban regeneration and sustainability)*. Franco Angeli editore.

Nesoti, A. (2007). *Parrocchia di S. Margherita V. M. in Vigonza. Cenni storici* (Parish of Santa Margherita V. M. in Vigonza. Historical notes), Parrocchia di S. Margherita V. M., Vigonza.

Pennacchi, A. (2011). *Fascio e martello. Viaggio per le città del Duce* (Beam and Hammer. Journey through the cities of the Duce), Editori Laterza, Rome-Bari.

Pietrogrande, E. (2011). *L'opera di Quirino De Giorgio (1937–1940): Architettura e classicismo nell'Italia dell'impero (The work of Quirino De Giorgio (1937–1940): Architecture and classicism in Italy during the age of the empire)*. Franco Angeli editore.

Rigamo, R. (1938). Vasta opera rigeneratrice nelle campagne padovane. Dal casone alla ridente casa rurale (Wide regenerative work in the Paduan countryside. From the the house of mud and straw to the smiling rural house), in "Padova", n. 10.

Schröder, U. (2017). *La Città degli spazi* (The City of Spaces), in "Lectiones. Riflessioni sull'architettura (Lectiones. Thoughts on architecture)", C. Orfeo, Clean edizioni, Naples.

Chapter 9
Walking in Vital Space: Timeline of Vitality in the Inner City of Urmia

Momen Foadmarashi and Francisco Serdoura

Abstract Public spaces play a pivotal role in urban life, acting as centres for social interaction, cultural exchange and economic activity. This study focuses on the Urmia Bazaar in Iran, a historic public space that exemplifies resilience and adaptability amidst extensive urban transformation. By employing space syntax analysis across three historical periods (1933, 1948 and 2023), the research investigates how spatial configurations, pedestrian flow patterns and accessibility dynamics have evolved over time and contributed to the vitality of the bazaar. The findings reveal that despite the introduction of modern grid systems, which altered the city's traditional layout, the bazaar has retained its role as a socio-economic and cultural nucleus. Metrics such as Normalized Angular Integration (NAIN) and Normalized Angular Choice (NACH) demonstrate how the bazaar sustains high levels of spatial integration and connectivity, enabling it to remain an attractor of pedestrian flows and a thriving hub of commerce and community engagement. The adaptability of its spatial configuration, combined with its embedded cultural significance, underscores the enduring relevance of traditional urban forms in modern contexts. This study also highlights the interplay between spatial design and urban vitality, signifying that well-connected and accessible spaces foster pedestrian movement and social interaction while supporting economic activity. The research identifies key challenges posed by modernization and offers strategies to enhance the vitality of historic public spaces. These include improving spatial connectivity through better pedestrian pathways, integrating sustainable design principles, promoting inclusive urban planning practices, and implementing cultural programming to reinforce the bazaar's socio-cultural role. The findings have significant implications for urban planners and policymakers seeking to balance heritage preservation with contemporary urban demands. By bridging theoretical insights with practical applications, this study contributes to

M. Foadmarashi (✉)
CIAUD, Research Centre for Architecture, Urbanism and Design, Lisbon School of Architecture, Universidade de Lisboa, Lisbon, Portugal
e-mail: mfoadmarashi@fa.ulisboa.pt

F. Serdoura
CIAUD, Research Centre for Architecture, Urbanism and Design, Lisbon School of Architecture, Universidade de Lisboa, Lisbon, Portugal
e-mail: fs@fa.ulisboa.pt

F. Naselli et al. (eds.), *Self-Sufficient Cities and Sustainable Habitats*, The Urban Book Series, https://doi.org/10.1007/978-3-032-10337-6_9

the broader discourse on public space vitality, offering a replicable framework for assessing and revitalizing historic spaces in rapidly urbanizing cities. In conclusion, the Urmia Bazaar serves as a testament to the resilience of traditional public spaces, providing valuable lessons for designing inclusive, vibrant and sustainable urban environments that honour cultural heritage while addressing modern needs.

Keywords Pedestrian flow · Vitality · Spatial analysis · Accessibility · Urmia · Bazaar · Iran

9.1 Introduction

The vitality of urban public spaces is a topic of ongoing debate, with researchers and practitioners alike recognizing their significance as the stage for social, cultural and economic interactions. Meanwhile public spaces reflect the design of a city fabric and, in turn, influence people's movement as well as traffic patterns.

They provide places for people to gather, socialize and relax. Indeed, walkable streets and squares as the main urban public spaces are particularly important for urban space vitality. They make it easier for people to walk and bike, which can improve public health and reduce traffic congestion. They can also create a livelier and welcoming atmosphere.

This paper discusses the importance of public space vitality and how it can be improved. It focuses on the Bazaar neighbourhood of Urmia, Iran, as an example of a pedestrian-friendly and vibrant public space. The paper uses the theory and methodology of space syntax to analyse the neighbourhood and identify its strengths and weaknesses. It concludes by proposing several strategies for improving the vitality of the Bazaar neighbourhood.

Urmia, a city with a history that spans over 5000 years, imbedded in north-west of Iran; has endured numerous conflicts and conquests. Despite being occupied, destroyed and rebuilt numerous times, its old fabric remains largely intact, with most structures dating back to the Qajar period. This historical structure (Fig. 9.1) offers a glimpse into Urmia's past and provide a fascinating contrast to the modern gridded streets introduced in the early twentieth century. While disrupting the city's traditional layout, these new grids revitalized public spaces by creating new opportunities for social interaction and economic activity.

9.2 Theoretical Framework

Urban public spaces serve as the heart of cities, fostering social interactions, economic vitality and a sense of community. Their vitality, or the level of activity, usage and social interaction they exhibit, is a critical factor in determining the overall liveability of a city. While numerous factors contribute to the vitality of urban public

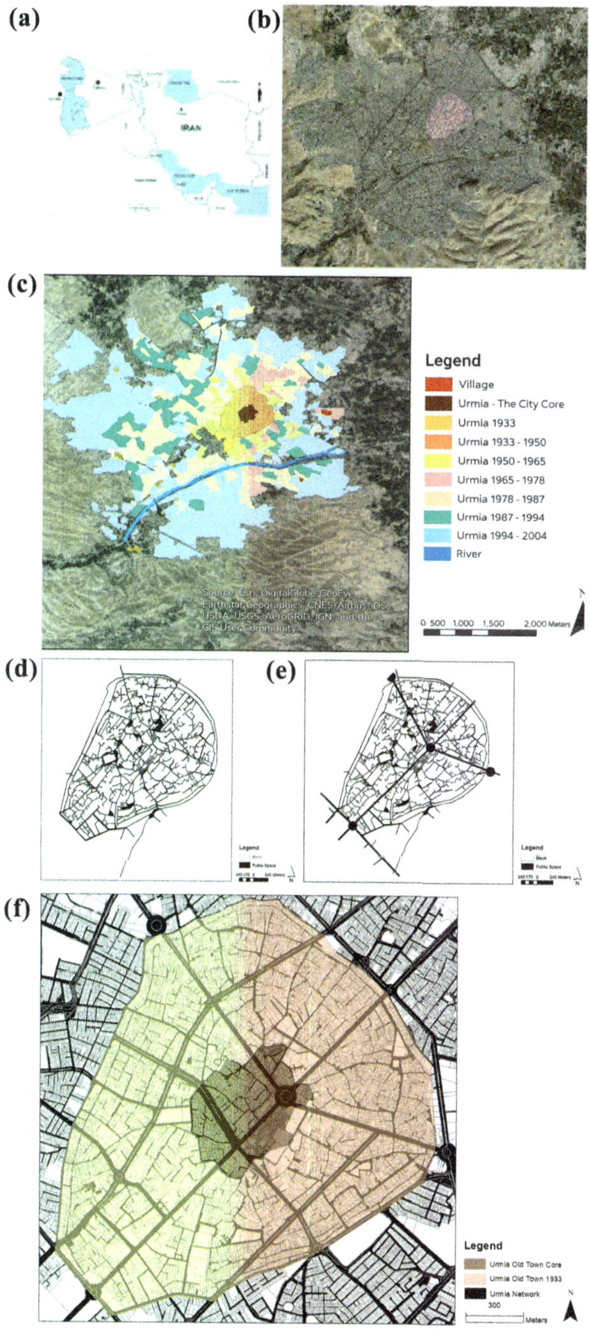

Fig. 9.1 Location of Urmia in Iran(**a**), Location of the inner city at present (**b**), Urmia city through the History (**c**) Urmia city in wall in 1933 (**d**), Urmia city after the first modern imposed streets 1948 (**e**), Urmia inner city in modern grid in 2018 (**f**)

spaces, walkability stands out as one of the key determinants. Walkability, the ease of using foot for transportation, is a fundamental aspect of vibrant public spaces. When people feel comfortable walking, they are more likely to frequent public spaces, leading to increased activity, interaction and overall vitality (Giles-Corti et al., 2016; Hillier, 1996; Madanipour, 2010). These investigations have consistently demonstrated that public spaces are not merely passive backdrops for human activity; rather, they actively shape and influence the patterns of human behaviour that occur within them.

This dynamic relationship is particularly evident in medieval cities/old fabrics/inner cities, where streets and squares were meticulously designed to support the diverse activities of residents (Gehl, 2011; Kostof, 1999). These spaces were not merely thoroughfares; they were hubs of social interaction, commerce and cultural exchange. The urban fabric of medieval cities fostered a vibrant public life, with people regularly gathering in the streets and squares to engage in a variety of activities.

The influence of public spaces on human behaviour has also been explored in the context of modernist urban planning. Jacobs (1961) recognized the importance of block size in shaping pedestrian activity. She argued that smaller blocks, with their shorter distances between destinations, encourage more walking and interaction among pedestrians. This emphasis on pedestrian-friendly designs has been further echoed in recent research, which highlights the positive impact of short blocks on social interactions and overall liveability (Penn et al., 1998).

Flowingly, walkability, a multifaceted concept encompassing both physical and social aspects, refers to the degree to which a built environment facilitates walking as a mode of transportation (Giles-Corti et al., 2016). It involves features such as continuous sidewalks, well-connected streets, safe crossings, access to public transit and amenities that encourage physical activity (Cervero & Kockelman, 1997). Studies have shown that walkable neighbourhoods are positively associated with social interactions and vitality, as they provide opportunities for people to gather, connect and engage in community activities (Abbaszadeh & Tamry, 2012; Foadmarashi and Serdoura, 2021a, b). This is attributed to the intelligibility, legibility, readability, safety and permeability of pedestrian spaces which make them inviting and conducive to social life.

Space syntax, a quantitative approach, plumbs the depths of urban spaces to decipher their intricate spatial arrangements and their latent vitality share. It employs metrics like connectivity, integration and choice to gauge how effortlessly individuals navigate and engage within these spaces (Hillier, 2009; Hillier & Hanson, 1984; Hillier et al., 1987). Research has unveiled those spaces imbued with high levels of connectivity and integration, facilitating multiple movement pathways and seamless access to essential destinations, tend to pulsate with vibrancy. Furthermore, spaces that espouse low levels of control, fostering a sense of open and liberating ambiance, are more likely to be perceived as welcoming and conducive to social interaction (Hillier, 2009; Hillier & Hanson, 1984). Subsequent analysis of permeability, using syntactic measures, can indicate the dynamism of a space, allowing planners to assess its ability to accommodate and encourage pedestrian movement. Hillier and Hanson (1984) classify permeability into two main aspects: integration, which measures how accessible a space is within the system, and choice which indicates how frequently a space is used as a shortcut between origins and destinations, influencing pedestrian movement patterns (Donegan et al., 2019; Hillier, 2009).

Meanwhile, space syntax analysis can shed light on the interconnectedness of various elements within an urban environment, revealing how people navigate and interact within its intricate spatial fabric. By identifying barriers to pedestrian accessibility, such as poor connectivity or limited access to key destinations, space syntax can empower urban design and planning strategies that promote more walkable environments. This valuable tool can help create pedestrian-friendly spaces that encourage social interactions and foster a vibrant urban spirit (Lin & Ban, 2017; Turner, 2001; van Nes & Yamu, 2021b; Ye et al., 2018).

9.3 Methodology

9.3.1 Theoretical Framework: Space Syntax

This study employs a quantitative approach to investigate the interplay between urban form and pedestrian movement within the Urmia Bazaar, Iran. A longitudinal analysis, spanning three distinct periods (1933, 1948 and 2023), is undertaken to understand how historical transformations of the urban fabric have influenced pedestrian flow patterns.

9.3.2 Methodology: Linear Representation and Space Syntax Analysis

This research is grounded in the theoretical framework of space syntax (Hillier & Hanson, 1984), a discipline that posits a strong correlation between the spatial configuration of the built environment and patterns of human movement. As Hillier and Hanson (1984) argue, space syntax emphasizes the inherent social logic embedded within urban form, asserting that the arrangement of streets, buildings and open spaces significantly influences pedestrian behaviour. This is particularly relevant in the context of historical bazaars like Urmia, where the spatial configuration has evolved over time, influencing pedestrian flow and the overall vitality of the space (Foadmarashi and Serdoura, 2021a).

Methodology: Linear Representation and Space Syntax Analysis

To capture the evolving spatial characteristics of the Urmia Bazaar, a linear representation of the city was generated for each historical period. This approach, involving the creation of axial maps, effectively simplifies the complex urban fabric into a network of interconnected lines, facilitating the analysis of changes in street patterns and connectivity over time (Hillier, 2009). This method aligns with the research conducted by Foadmarashi and Serdoura (2021b), where they analysed the Urmia Bazaar's spatial configuration and its impact on pedestrian movement across different time periods.

Space syntax analysis was employed to quantify the spatial properties of the Urmia Bazaar. This methodology offers several advantages for this study:

- **Handling Complexity**: Space syntax effectively analyses the intricate network of streets and passages characteristic of traditional bazaars (Turner, 2001), capturing the nuanced spatial relationships within these complex environments, as demonstrated in the analysis of the Urmia Bazaar by Foadmarashi and Serdoura (2021a).
- **Quantitative Measurement**: It provides a set of quantitative measures, such as integration, choice and depth, that allow for rigorous analysis of how easily pedestrians can navigate and access different parts of the bazaar (Hillier et al., 1993).
- **Link to Pedestrian Behaviour**: Space syntax theory suggests that spaces with higher integration and connectivity tend to attract more pedestrian flow (Hillier, 2009). This study investigates whether this theoretical relationship holds true within the dynamic context of the Urmia Bazaar, building upon the findings of Foadmarashi and Serdoura (2021b) that highlighted the relationship between spatial configuration and pedestrian movement in the Bazaar.

9.3.3 Selection of Space Syntax Measures

This study focuses on two key space syntax measures: Normalized Angular Integration (NAIN) and Normalized Angular Choice (NACH), introduced by Hillier, et al. (2012). These metrics provide a more nuanced understanding of pedestrian movement and accessibility within urban spaces:

- **Normalized Angular Integration (NAIN)**: This metric quantifies the degree to which a street segment is integrated within the overall street network. It reflects the ease with which a particular location can be reached from other parts of the urban fabric. Normalizing the integration value allows for comparisons between urban fabrics of different sizes and complexities.
- **Normalized Angular Choice (NACH)**: This measure indicates the probability of a street segment being used in pedestrian routes. It reflects the range of movement choices available at a particular location. This metric helps to identify streets/ public spaces that act as important thoroughfares or 'attractors' of pedestrian movement and as centrality in urban area.

NAIN and NACH were chosen for their ability to provide a comprehensive assessment of both local and global accessibility within the Urmia Bazaar. By analysing these measures across different time periods, the research aims to identify how changes in the urban fabric have influenced pedestrian movement/walking patterns and accessibility within this vital public space. These measures can provide us more classifications classification:

- **High Integration/High Choice**: These areas are likely to be highly accessible and attractive to pedestrians, offering a wide range of movement options and good connectivity within the network. They may represent key thoroughfares, commercial centres or popular gathering places.
- **High Integration/Low Choice**: These areas are highly accessible but offer limited movement choices. They may represent cul-de-sacs, dead-ends or areas with limited connectivity beyond the immediate vicinity.
- **Low Integration/High Choice**: These areas are less accessible but offer a variety of movement options. They may represent peripheral areas with limited connections to the main street network but with internal connections and a diverse range of destinations.
- **Low Integration/Low Choice**: These areas are both less accessible and offer limited movement choices. They may represent isolated areas, dead-ends or areas with poor connectivity to the rest of the urban network.

By analysing NAIN and NACH values and classifying street segments based on these metrics, this study aims to identify areas of high and low accessibility, connectivity, and pedestrian activity within the Urmia Bazaar. This classification system will provide a more nuanced understanding of the spatial dynamics of pedestrian movement and inform urban planning strategies for enhancing the vitality of the Bazaar, aligning with the recommendations proposed by Foadmarashi and Serdoura (2021a) for improving the vitality of public spaces in Urmia.

9.3.4 Data Collection and Analysis

The research process involved the following steps:

1. **Map Preparation**: Historical maps from 1933 and 1948 were obtained from Cultural Heritage, Tourism and Handicraft Organization of the West Azarbaijan, digitalized and georeferenced using Auto Cad and ArcGIS software. The 2023 street network was extracted from OpenStreetMap (OSM) data. Also, has used the last approved map of the city Of Urmia as detailed plan of the city. All maps were digitized to create vector representations of the streets and paths.
2. **Linear Representation**: Axial maps were generated for each period by identifying the fewest and longest lines that represent the dominant lines of sight and movement within the urban fabric. Meanwhile, Segment map prepared for each period to analysis the topological and angular measures for the study. However, for the latest period we used the OSM centrelines as segments.
3. **Data Refinement**: The axial maps were overlaid on orthophoto maps from Google Earth to verify the accuracy of the street network. Field surveys conducted in 2018 and 2022 were used to identify any discrepancies between the maps and

the actual built environment, such as changes in street widths or the addition of new pedestrian pathways. These observations were used to refine and update the axial maps or refined centrelines as segment map.

4. **Space Syntax Analysis**: The refined axial maps were imported into DepthmapX network Version 0.35 and QGIS version 3.28 with Space Syntax Toolkit Plug-in software to calculate space syntax metrics, including NAIN and NACH for each street segment.
5. **Statistical Analysis**: To further investigate the relationship between space syntax measures and pedestrian flow, statistical linear regression analysis was conducted. This statistical method was employed to quantify the strength and direction of the relationship between connectivity (as measured by, for example, global integration) and pedestrian flow surrogates such as NACH and NAIN. Linear regression models were developed to predict NACH and NAIN values based on connectivity measures, allowing for a quantitative assessment of the extent to which connectivity influences these key space syntax indicators.

9.3.5 Limitations

While this methodology provides valuable insights, it's important to acknowledge its limitations. The accuracy of the analysis depends on the quality of the historical maps, which may contain inaccuracies or omissions. Additionally, the linear representation might oversimplify the complexity of the urban fabric, and space syntax may not fully capture all factors influencing pedestrian movement, such as visual attractiveness, social dynamics and socio-economic factors.

9.4 Case Study Area

Urmia, situated on a plain 17 kms west of Lake Urmia at an elevation of 1360 m above sea level, occupies a geographical area bounded by 45°00' to 45°06' longitude and 37°30' to 37°36' latitude. While its precise origins remain shrouded in earlier history. Although the history of the city routed to ancient times regarding to archaeological excavation, the earliest printed documented evidence of the city, dating back to the fourth century, attests to the presence of a grand mosque, a bustling bazaar and a fortified city wall, suggesting a degree of urban development and civic infrastructure even at that early stage (Abasszade, 2011).

Jackson's travelogue (1906) provides a glimpse into the historical defences of Urmia, which included a five-kilometre wall with seven gates and a network of moats strategically placed to repel attackers.

Nestled within the heart of Urmia's inner city, the Urmia Bazaar stands as a testament to the city's rich history and vibrant spirit. This expansive urban hub, encompassing residential quarters, lively markets and covered passages, has long served as a central crossroads for the city's inhabitants, offering a tapestry of cultural experiences, diverse activities and a dynamic mix of land uses. Habibi (2000) and Soltanzadeh (2011) described the Bazaar as the 'biggest and first-born covered public space of the city' and 'the social, cultural and economic centre of the city', respectively. The Bazaar's central location and rich history have made it a vital component of Urmia's identity, attracting people from all walks of life (Foadmarashi and Serdoura, 2021a, 2021b).

The Bazaar, Urmia's oldest and grandest enclosed public space, has traditionally maintained close ties with its surrounding neighbourhoods, fostering a sense of coherence and shared identity. More than just a passage through the city's fabric, the Bazaar has played a key role as the social, cultural and economic heart of Urmia (Fig. 9.2).

Over the course of the twentieth century, Urmia's rapid modernization and urban planning efforts brought about significant transformations to the city's layout. The imposition of rigid grids extended into the inner city, altering the traditional patterns of streets, squares and public spaces, including the Bazaar's enclosed passages. These alterations, while driven by modernization goals, also reshaped the social and spatial dynamics within the city, influencing the way Urmia's inhabitants interacted with their urban environment.

Fig. 9.2 Urmia Bazaar, General view and how it looks like

9.5 Analysis and Results

Urmia, a monocentric city with a densely populated inner core, boasts the bazaar, one of its oldest and most vibrant commercial centres. This labyrinthine network of interconnected spaces serves as a crucial economic hub, fostering connections between the city's diverse neighbourhoods. Spatial analysis using axial lines, a technique that visualizes the city's major pathways, reveals the bazaar's enduring prominence as a highly trafficked public space. Segment maps, representing the urban fabric's linear structure, highlight the bazaar's role in shaping the city's compact layout, particularly in the central and inner districts, where narrow streets and dead-ends are prevalent (Fig. 9.3).

Figure 9.3 offers a compelling visual representation of pedestrian mobility patterns around the historic Urmia Bazaar, demonstrating how the bazaar's presence has influenced movement patterns both before and after the municipality's introduction of a new grid system.

It is evident that individuals prioritize direct routes in their daily commutes, often opting to utilize the covered bazaar as a time-efficient means of accessing the city centre. Despite the introduction of new streets, which has altered the distribution of walking spaces from a centralized to a linear pattern, the bazaar continues to serve as an integral component of the old town's urban fabric (Figs. 9.4).

The metrics of NACH and NAIN, which serve as indicators of pedestrian accessibility and route choice probabilities, reveal a notable increase over time, underscoring the city centre enduring significance. Despite the expansion of the city's grid system in recent years, overall connectivity levels have remained remarkably consistent, both in the city centre and across the entire city. This stability suggests that the city's accessibility has not undergone significant alterations despite the infrastructure developments. However, the increasing values for integration and choice, particularly in the city centre, point to a growing utilization of its spaces, indicating that the centre's appeal remains strong among pedestrians. This highlights the city centre's ability to adapt to evolving needs while preserving its central role in the urban fabric.

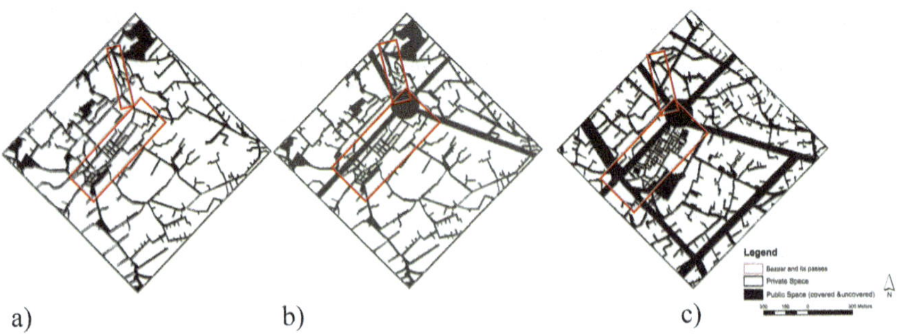

Fig. 9.3 Changes in the fabric of Urmia Bazaar neighbourhood in 1933 (**a**), 1948 (**b**) and 2023 (**c**)

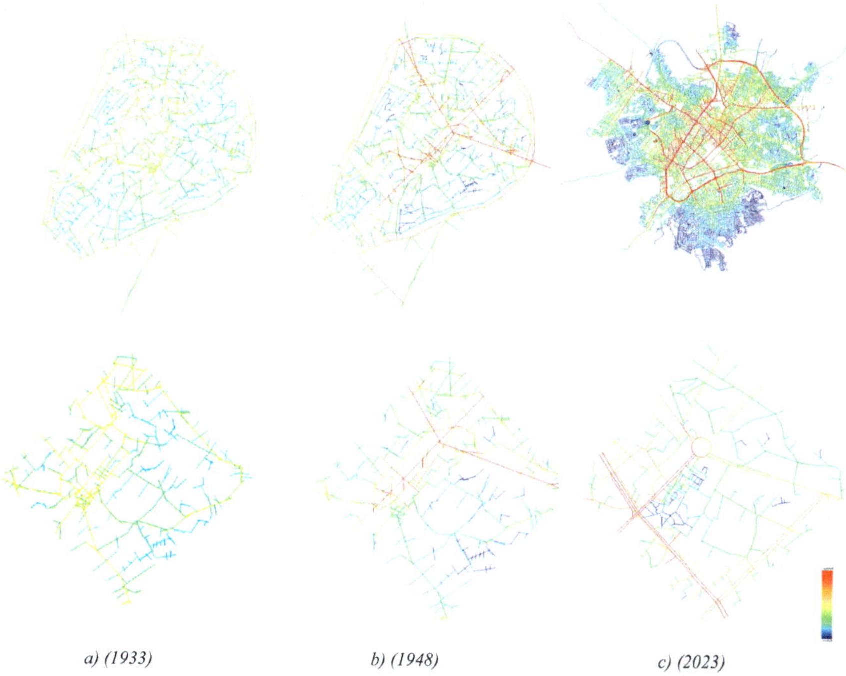

Fig. 9.4 Comparison of the NAIN in three different periods in Urmia inner city and Bazaar block

The enduring significance of Urmia's bazaar lies in its ability to facilitate efficient movement and economic exchange, while simultaneously preserving the city's rich cultural heritage and architectural legacy. The bazaar stands as a testament to the enduring power of traditional urban forms in adapting to evolving needs and maintaining their role as vibrant hubs of community life.

This Linear representation of the city in three period can also provide a more coherence and more integrated spaces in the city centre as well as the whole city. Although modernization movements affected the coherence of the fabric in a dramatic way by cutting the city, as it was appeared in Table 9.1, the centrality of the spaces in the city and their accessibility was not changed a lot in time and even was not that much different and has more better distribution regarding concerning the time as the R is decreasing in time from 0.35 in the study area reached to 0.14 which was in the same streamline of the path in the city.

Table 9.1 NACH versus NAIN in linear representation of Urmia inner city versus Urmia Bazaar Block study area in three periods

Year	Level	NACH		NAIN		Connectivity	
		whole city	Study area	whole city	Study area	whole city	Study area
1933	Max	***1.4623***	***1.4583***	1.1302	1.1302	***8.0***	6.0
	Min	0.000	0.000	0.3825	0.5009	1.0	1.0
	Mean	1.4623	0.7872	0.8206	0.8432	4.1639	4.3728
1948	Max	1.6293	1.6293	1.7350	1.7244	8	6
	Min	0.000	0.000	0.4958	0.5294	1.0	1.0
	Mean	0.7324	0.8200	1.0596	1.1057	4.2150	4.4718
2023	Max	***3.1029***	1.3712	***6.6956***	1.6903	7	***7***
	Min	0	0.000	0.2652	0.5575	1	1
	Mean	0.8368	0.8828	0.7492	0.7989	3.202	3.0708

9.6 Discussion

The vitality of urban public spaces, as exemplified by the Urmia Bazaar, is inherently tied to their spatial configuration, socio-cultural significance and adaptability to evolving urban dynamics. This study, employing space syntax analysis, underscores the bazaar's resilience and highlights its role as a pivotal socio-economic and cultural node within Urmia's urban fabric. The discussion critically interprets the findings, situates them within broader theoretical frameworks and considers their implications for urban planning and policy.

The results demonstrate that the Urmia Bazaar retains high levels of spatial integration and connectivity across all three analysed periods (1933, 1948 and 2023). Metrics such as Normalized Angular Integration (NAIN) and Normalized Angular Choice (NACH) reveal the enduring accessibility and navigability of the bazaar, even as modern grid structures were superimposed on the city fabric. These findings align with Hillier's (2009) assertion that spaces with higher integration values naturally attract greater pedestrian activity, reinforcing their socio-economic vibrancy (Fig. 9.5).

While modernization efforts introduced linear patterns and new infrastructure, the bazaar's core spatial logic remained intact. This resilience can be attributed to its historical configuration, which prioritizes compact, interconnected pathways. The findings echo Penn et al. (1998), who argue that well-connected urban spaces are better equipped to sustain vibrant pedestrian flows. The bazaar's adaptability underscores the capacity of traditional spaces to coexist with modern planning interventions without compromising their essential functionality.

The Urmia Bazaar exemplifies the dual role of traditional marketplaces as both economic engines and cultural repositories. As noted by Madanipour (2010), bazaars in Persian and Islamic cities have historically functioned as spaces of cultural

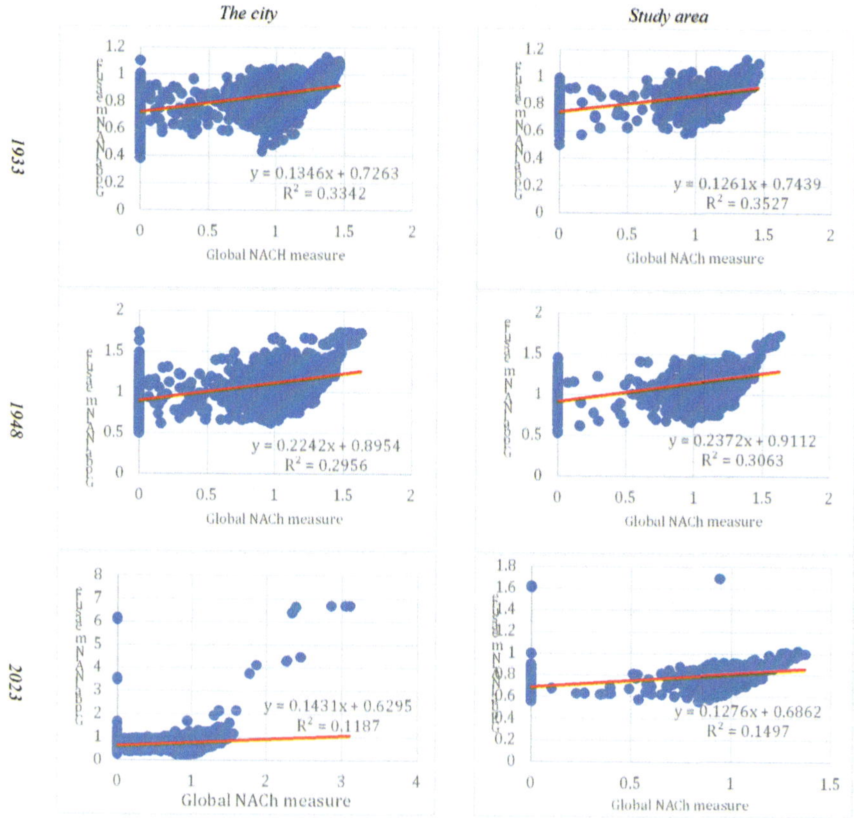

Fig. 9.5 Scatter plots comparing the Global NAIN measure to the Global NACH measure across different contexts and years in Urmia inner city versus Urmia bazaar block study area in 1933,1948, 2023

exchange, social interaction and economic activity. The covered passages and intricate network of pathways in the Urmia Bazaar not only support commerce but also foster community cohesion and cultural continuity.

The study provides actionable insights for enhancing the vitality of public spaces, with specific recommendations for the Urmia Bazaar and similar contexts:

- **Enhancing Spatial Connectivity**: Strengthening pedestrian linkages within the bazaar and to adjacent neighbourhoods will improve its accessibility, as suggested by Giles-Corti et al. (2016). Strategic integration of public transit stops near key entrances would further support equitable access.
- **Promoting Cultural Programming**: Festivals, artisan workshops and guided heritage tours can enhance the bazaar's role as a cultural hub. Such initiatives align with Gehl and Svarre's (2013) emphasis on activating public spaces through community engagement.

- **Inclusive Design**: Ensuring pathways and facilities are universally accessible fosters equity and inclusivity, reflecting Abbaszadeh and Tamry's (2012) advocacy for pedestrian-friendly spaces that cater to diverse users.
- **Integrating Sustainability Measures**: Introducing green infrastructure, such as small parks or eco-friendly materials, and implementing sustainable practices, such as waste reduction systems, will align with urban sustainability goals (Cervero & Kockelman, 1997).

These strategies demonstrate that the vitality of public spaces is contingent not only on their spatial qualities but also on their ability to adapt to contemporary socio-economic and environmental needs.

The findings contribute to the broader discourse on urban vitality by reinforcing and extending existing theories. Jacobs' (1961) emphasis on walkability and mixed-use diversity is reflected in the bazaar's compact layout, which encourages pedestrian interaction and economic activity. Similarly, Hillier's (2009) space syntax framework is validated by the strong correlation between integration values and pedestrian movement patterns observed in the Urmia Bazaar.

However, this study advances these theories by illustrating the adaptability of historical urban cores to modern interventions. Unlike the prevalent narrative that modernization dilutes the essence of traditional spaces, the Urmia Bazaar demonstrates how these spaces can maintain their socio-cultural and functional significance through strategic adaptation.

While the findings provide valuable insights, certain limitations warrant attention. The reliance on space syntax metrics, while robust, does not capture intangible socio-cultural dynamics such as the behavioural motivations of visitors or the informal economies operating within the bazaar. Future research could integrate ethnographic methods, demographic studies and real-time pedestrian tracking to complement the quantitative spatial analysis (Donegan et al., 2019).

Furthermore, this study focuses exclusively on the Urmia Bazaar. Comparative analyses of other traditional marketplaces across diverse cultural and geographic contexts could offer a more comprehensive understanding of the resilience of public spaces. Exploring the role of emerging technologies, such as digital navigation tools or smart city systems, in enhancing accessibility and vitality also presents an avenue for future research.

The Urmia Bazaar serves as a microcosm of the challenges and opportunities facing historical public spaces worldwide. Its ability to retain vitality amidst modernization offers a compelling model for urban planners aiming to balance heritage conservation with contemporary urban development. Comparative insights from similar spaces, such as London's Borough Market or Tehran's Grand Bazaar, highlight universal principles of adaptability while underscoring the unique socio-cultural dynamics that shape each space.

By demonstrating the interplay between spatial configuration, cultural continuity and modern infrastructure, this study emphasizes the potential of traditional public spaces to thrive in evolving urban landscapes.

The Urmia Bazaar exemplifies the resilience and adaptability of traditional public spaces, offering valuable lessons for urban planning and policy. By prioritizing spatial connectivity, socio-cultural integration and sustainable practices, planners can ensure the vitality of such spaces in the face of modernization. This study contributes to the broader understanding of urban vitality, emphasizing the importance of preserving the unique character of public spaces while adapting them to meet the needs of future generations.

9.7 Conclusions

The vitality of urban public spaces lies at the heart of their ability to adapt to evolving socio-economic, cultural and spatial dynamics. This study on the Urmia Bazaar highlights the enduring significance of traditional marketplaces as vibrant hubs of social interaction, commerce and cultural exchange, even amidst significant urban transformations. By employing space syntax analysis, the research has demonstrated how spatial connectivity, accessibility and pedestrian movement patterns contribute to sustaining the bazaar's vitality over time.

To build on these findings, a comprehensive framework of strategic interventions is essential to safeguard and enhance the bazaar's role in the urban fabric. First, improving spatial connectivity is crucial. By developing additional pedestrian pathways and enhancing integration with surrounding neighbourhoods, the bazaar can attract broader and more diverse users while fostering a seamless flow of movement. Additionally, integrating public transit routes directly with the bazaar's entrances will strengthen its accessibility, enabling equitable access for residents from across the city.

Second, the adoption of inclusive urban design principles is necessary to ensure the space remains welcoming and accessible to individuals of all abilities. This inclusivity reinforces the social and communal dimensions of the bazaar as a public space. Furthermore, cultural programming, such as regular festivals, heritage tours and community-led workshops, will celebrate the bazaar's unique identity and deepen its connection to both local and global audiences.

To sustain economic vitality, economic diversification and support for traditional craftspeople must be prioritized. Encouraging mixed-use spaces, while providing targeted incentives for artisanal trades, ensures the preservation of cultural heritage alongside the introduction of contemporary urban uses. Simultaneously, integrating green infrastructure and sustainable practices, such as waste reduction, renewable energy installations and pedestrian-friendly landscaping, will align the bazaar with modern sustainability goals, improving environmental quality and user experience.

Preserving the bazaar's historical and architectural essence is also critical. This can be achieved through the implementation of heritage conservation policies that balance the need for modern functionality with the protection of the bazaar's cultural legacy. Such measures will ensure the space retains its unique character while adapting to contemporary needs.

Effective governance is vital to the successful implementation of these strategies. A community-centric management model, engaging local stakeholders, vendors and planners, will ensure interventions are rooted in the needs and aspirations of the community. Moreover, the deployment of real-time monitoring systems to track pedestrian flows, space usage and visitor satisfaction will enable dynamic, data-driven adjustments to maintain vitality. Long-term resilience can be further secured through periodic evaluations and strategic planning to respond proactively to emerging challenges and opportunities.

In conclusion, the Urmia Bazaar stands as a testament to the adaptability and resilience of traditional public spaces in navigating the challenges of modernization. By addressing its spatial, cultural and economic dimensions holistically, the bazaar can continue to thrive as a dynamic and inclusive urban hub. These strategies not only reinforce its role in the urban fabric of Urmia but also offer valuable insights for safeguarding the vitality of public spaces in cities worldwide. Through careful planning, preservation and innovation, public spaces like the Urmia Bazaar can sustain their socio-economic and cultural relevance for future generations.

Acknowledgements To the Portuguese Foundation of Sciences and Technology (FCT) for the Ph.D. scholarship of the first author, since this work is financed by national funds through FCT—Fundação para a Ciência e a Tecnologia, I.P., under the Ph.D. scholarship with the reference SFRH/BD/147266/2019.

This work is financed by national funds through FCT—Fundação para a Ciência e a Tecnologia, I.P., under the Strategic Project with the references UIDB/04008/2020 and UIDP/04008/2020.

References

Abasszade, M. (2011). Studying the special arrangement of the historic city center of theUrmia. Symposium on the Historical, Religious, and Cultural City of Urmia, Urmia.

Abbaszadeh, S., & Tamry, S. (2012). Analysis of factors affecting the improvement of Pedestrian walkway spatial quality and Pedestrian-oriented Spaces, in order to Increase the social interactions Level of People (the case study: Tarbiat & Valiasr axis, Tabriz metropolitan). *Motaleate Shahri, 1*(4), 95–104.

Cervero, R., & Kockelman, K. (1997). Travel demand and the 3Ds: Density, diversity, and design. *Transportation Research Part d: Transport and Environment, 2*(3), 199–219. https://doi.org/10.1016/S1361-9209(97)00009-6

Donegan, L., Silveira, J. A. R. da, & Silva, G. J. A. da. (2019). Under and over: Location, uses and discontinuities in a centrally located neighbourhood in João Pessoa city reflecting current urban planning effects. *Urbe. Revista Brasileira de Gestão Urbana, 11*. https://doi.org/10.1590/2175-3369.011.e20180123

FoadMarashi, M., & Serdoura, F. (2021). Public space use: An approach to livability in Urmia Bazaar, Iran. In S. Eloy, D. Leite Viana, F. Morais, & J. Vieira Vaz (Eds.), *Formal Methods in Architecture* (pp. 231–240). Springer International Publishing. https://doi.org/10.1007/978-3-030-57509-0_21

Foadmarashi, M., & Serdoura, F. (2021a). Vitality in time: An approach to the use of Urmia Bazaar. In *Proceedings Book of the Fourth International Conference on Architecture and Urban Design*, vol 1 (pp. 131–140). http://dspace.epoka.edu.al/handle/1/2248

FoadMarashi, M., & Serdoura, F. (2022b). The effect of new urban design action on promoting the public space vitality in the inner city. In *Annual Conference Proceedings of the XXVIII International Seminar on Urban Form "Urban Form and the Sustainable and Prosperous City"* (pp. 1253–1260).

Gehl, J. (2011). *Life between buildings: Using public space.* Island Press. ISBN 1-61091-023-0.

Gehl, J., & Svarre, B. (2013). *How to study public life.* Island Press. ISBN 978-1-61091-525-0.

Giles-Corti, B., Vernez-Moudon, A., Reis, R., Turrell, G., Dannenberg, A. L., Badland, H., Foster, S., Lowe, M., Sallis, J. F., Stevenson, M., & Owen, N. (2016). City planning and population health: A global challenge. *The Lancet, 388*(10062), 2912–2924. https://doi.org/10.1016/S0140-6736(16)30066-6

Habibi, S. M. (2000). *De la Cite de la Ville; A historical analysis of the concept of the City and its physical appearance.* (1st ed.). University of Tehran Press.

Hillier, B. (1996). Cities as movement economies. *Urban Design International, 1*(1), Article 1.

Hillier, B. (2009). Spatial sustainability in cities: Organic patterns and sustainable forms [Proceedings paper]. In: D. Koch, L. Marcus, & J. Steen (Eds.), *Proceedings of the 7th International Space Syntax Symposium.* (p. 1). Royal Institute of Technology (KTH): Stockholm, Sweden (2009). Royal Institute of Technology (KTH). http://www.sss7.org/Proceedings_list.html

Hillier, B., Burdett, R., Peponis, J., & Penn, A. (1987). Creating life: Or, does architecture determine anything? *Architecture Et Comportement/architecture and Behaviour, 3*(3), 233–250.

Hillier, B., & Hanson, J. (1984). *The social logic of space.* Cambridge University Press. https://doi.org/10.1017/CBO9780511597237

Hillier, B., Penn, A., Hanson, J., Grajewski, T., & Xu, J. (1993). Natural movement: Or, configuration and attraction in urban pedestrian movement. *Environment and Planning B: Planning and Design, 20*(1), 29–66. https://doi.org/10.1068/b200029

Hillier, W. R. G., Yang, T., & Turner, A. (2012). Normalising least angle choice in Depthmap—and how it opens up new perspectives on the global and local analysis of city space. *Journal of Space Syntax, 3,* 155–193.

Jackson, A. V. W. (Abraham V. W. (with Robarts—University of Toronto). (1906). Persia past and present: A book of travel and research. Macmillan, New York . http://archive.org/details/persiapastpresen00jackuoft

Jacobs, J. (1961). *The death and life of great American Cities* (1st ed.). Random House.

Kostof, S. (1999). *The City assembled by Kostof, Spiro* (1st ed.). Bulfinch. ISBN 978-0-8212-2599-8.

Lin, J., & Ban, Y. (2017). Comparative analysis on topological structures of urban street networks. *ISPRS International Journal of Geo-Information, 6*(10), Article 10. https://doi.org/10.3390/ijgi6100295

Madanipour, A. (Ed.). (2010). *Whose public space?: International case studies in urban design and development.* Routledge.

Penn, A., Hillier, B., Banister, D., & Xu, J. (1998). Configurational modelling of urban movement networks. *Environment and Planning B: Planning and Design, 25*(1), 59–84. https://doi.org/10.1068/b250059

Soltanzadeh, H. (2011). *A brief history of the city and urbanization in Iran* (ancient Era to1976) (1st ed.). Chahar Tagh.

Turner, A. (2001). Depthmap: A program to perform visibility graph analysis. In *The Third International Space Syntax Symposium,* Georgia Technological Institute.

van Nes, A., & Yamu, C. (2021a). *Introduction to space syntax in urban studies.* Springer International Publishing. https://doi.org/10.1007/978-3-030-59140-3

van Nes, A., & Yamu, C. (2021b). Orientation and wayfinding: Measuring visibility. In A. van Nes, & C. Yamu (Eds.), *Introduction to Space Syntax in Urban Studies* (pp. 87–111). Springer International Publishing. https://doi.org/10.1007/978-3-030-59140-3_3

Ye, Y., Li, D., & Liu, X. (2018). How block density and typology affect urban vitality: An exploratory analysis in Shenzhen China. *Urban Geography, 39*(4), 631–652. https://doi.org/10.1080/02723638.2017.1381536

Chapter 10
How Walking-Streets Promoting Sightseeing: Effect of Walking Routes on the Tourist Places of Hamadan Inner City

Solmaz Asadpour, Momen Foadmarashi, and Francisco Serdoura

Abstract Over the past decades, walkable streets have emerged as a pivotal focus in urban design, recognized for their ability to enhance pedestrian mobility and support urban tourism. This study examines the impact of well-designed pedestrian pathways on tourism in Hamadan's inner city, a location renowned for its rich cultural and historical significance. By focusing on a non-European context, this research expands the discourse on walkable streets, demonstrating their potential to improve accessibility and elevate the appeal of historical tourist destinations. Hamadan's unique urban fabric, shaped by its role as an ancient cultural hub and a modernizing city, provides a distinctive lens to explore these dynamics. Using a mixed methods approach, the study integrates spatial analysis via Space Syntax techniques, field surveys, interviews and behavioural tracking to evaluate pedestrian flow, spatial perception and socio-economic outcomes. The findings reveal that pedestrian pathway within mid-range radii (1200–3600 m) are most effective in supporting tourist accessibility and activity, aligning with natural pedestrian movement patterns. Furthermore, the study highlights the role of walkable streets in promoting sustainability by reducing reliance on motorized transport and preserving cultural heritage. The present research aims to contribute to the global discourse on sustainable urban development by showcasing how pedestrian-friendly designs in non-Western historical cities can align with broader socio-economic and environmental goals. The insights gained from Hamadan's case study offer valuable lessons for integrating cultural heritage and pedestrian infrastructure into urban planning frameworks worldwide.

S. Asadpour (✉) · M. Foadmarashi · F. Serdoura
CIAUD, Research Centre for Architecture, Urbanism and Design, Lisbon School of Architecture, Universidade de Lisboa, Lisbon, Portugal
e-mail: solmaz.asadpour@edu.ulisboa.pt

M. Foadmarashi
e-mail: mfoadmarashi@fa.ulisboa.pt

F. Serdoura
e-mail: fs@fa.ulisboa.pt

F. Naselli et al. (eds.), *Self-Sufficient Cities and Sustainable Habitats*, The Urban Book Series, https://doi.org/10.1007/978-3-032-10337-6_10

Keywords Pedestrian flow · Sustainable tourism · Spatial analysis · Accessibility · Hamadan · Iran

10.1 Introduction

Walking is one of the most natural and fundamental forms of human movement, deeply intertwined with the history of urban life. Beyond its practical utility as a mode of transportation, walking serves as a powerful medium for individuals to perceive spatial identity, establish a sense of belonging and engage with the cultural, social and historical narratives of cities. Urban environments, when explored on foot, reveal their layered complexities, offering insights into their essence. However, modern urban planning practices have often prioritized vehicular transportation, leading to the fragmentation of urban spaces, a decline in walkability and the erosion of social, cultural and aesthetic values.

Pedestrian spaces, once vibrant centres of community interaction, have been overshadowed, resulting in diminished urban vitality and reduced engagement with public spaces. In response to these challenges, pedestrian-oriented urban design gained prominence in European cities during the mid-twentieth century. Post-war reconstruction efforts in cities like Stockholm and Rotterdam introduced pioneering pedestrianization strategies, emphasizing the preservation of historical identity and the promotion of urban liveability (Mehdizadeh, 2000). These initiatives established a foundation for sustainable and walkable urban environments. While these efforts have been widely studied in European contexts, their application in non-Western cities remains underexplored.

Hamadan, one of Iran's oldest cities, offers a unique opportunity to adapt and examine global pedestrianization principles in a culturally and historically distinct context. With a history spanning the Median era, the Islamic period, and its significance along the Silk Road, Hamadan boasts a rich urban fabric shaped by centuries of cultural exchange and modernization. Renowned for its historical landmarks and vibrant cultural attractions, the city attracts a diverse mix of domestic and international tourists, making it an ideal case study for investigating the interplay between pedestrian infrastructure and urban tourism.

This study aims to evaluate how walkable streets can enhance Hamadan's urban environment by improving accessibility, preserving cultural heritage and supporting sustainable tourism. Using the city's inner core as the focal point, the research examines the role of pedestrian infrastructure in bridging the gap between urban planning, socio-economic development and environmental sustainability. Hamadan's context provides fertile ground for exploring how well-designed pedestrian networks can reflect the cultural and geographical diversity of non-Western cities while contributing to the broader discourse on walkability and sustainable urban design.

10.2 Methodology

Mixed methods research has become an essential approach in fields such as tourism and urban studies, offering a robust framework to explore complex phenomena. By integrating qualitative and quantitative methodologies, researchers can address diverse dimensions of their study, ensuring both breadth and depth in their findings. This flexibility is particularly valuable in understanding multi-dimensional fields like tourism, where the interplay of cultural, social and economic factors demands diverse methodological perspectives.

In tourism research, the application of mixed methods enables a dual exploration of general patterns and specific contexts. Quantitative approaches, such as surveys and statistical analysis, capture large-scale trends like tourist satisfaction or pedestrian flow, while qualitative methods, such as interviews or ethnographic observations, delve deeper into visitor experiences and behaviours. Halcomb (2019) underscores that the strength of mixed methods lies in its ability to creatively combine these components, tailoring research designs to specific objectives. This study applies a mixed methods approach to examine the impact of walkable streets on tourism within Hamadan's inner city, integrating spatial analysis, field studies and statistical techniques to analyse pedestrian movement, urban infrastructure and socio-economic interactions (Halcomb, 2019). The combination of these methodologies ensures a comprehensive understanding of how pedestrian infrastructure influences tourism dynamics and supports sustainable urban development.

10.2.1 Spatial Analysis Using Space Syntax Theory

The spatial configuration of a city plays a critical role in shaping pedestrian behaviour and accessibility. Building on Space Syntax theory (Hillier & Hanson, 1987; van Nes et al., 2021), this study evaluates Integration and Choice metrics to understand connectivity and movement within Hamadan's historical urban fabric:

- **Integration** measures the accessibility of a space within the broader network, highlighting areas that naturally attract pedestrian flow due to their centrality.
- **Choice** assesses the likelihood that a street segment serves as a route between origins and destinations, indicating its importance as a connector. (Hillier et al., 1987)

Using road centrelines from OpenStreetMap (OSM), Angular Segment Analysis (ASA) was performed with the Space Syntax Toolkit in QGIS. The analysis was conducted across multiple radii (400, 800, 1200, 1600, 2000, 2400, 2800, 3200, 3600 and 4000 m) to capture pedestrian behaviours at varying scales. This revealed that mid-range radii (2000–3600 m) were critical for pedestrian activity, aligning with natural movement patterns and enhancing access to key tourist landmarks.

10.2.2 Field Studies for Validation

To complement the spatial analysis, field studies were conducted in Hamadan's central square and key pedestrian routes to validate findings:

- **Observation and Tracing**: Pedestrian and tourist movements were tracked to identify dynamic and static spaces. The central square emerged as a hub of pedestrian activity, while certain streets showed lower utilization, reflecting disparities in infrastructure effectiveness.
- **Behavioural Analysis**: Surveys and direct observations captured pedestrian interactions with public spaces, revealing patterns of foot traffic near historical landmarks and the influence of urban design on visitor behaviour.

10.2.3 Data Integration and Statistical Techniques

The findings from spatial analysis and field studies were synthesized to provide a holistic understanding of pedestrian behaviour. Urban land use maps were incorporated to align spatial data with functional attributes, illustrating how pedestrian pathways intersect with key attractions and urban features. To substantiate the findings further, linear regression analysis was employed to quantify the relationship between pedestrian flow, urban design variables and socio-economic benefits. Key variables included pedestrian volume, distance to attractions and integration values, offering insights into how walkability influences tourism dynamics.

10.2.4 Rationale for Methodology

Mixed methods research enables the integration of diverse data types, making it particularly effective for exploring the intersection of walkability and tourism in Hamadan's historical context. Spatial analysis identifies critical routes and connectivity patterns, while field observations validate these findings in real-world settings. Regression analysis provides quantitative support for the socio-economic implications of pedestrian infrastructure. By combining these approaches, this study offers actionable insights for urban planners and policymakers aiming to enhance tourism, sustainability and accessibility in Hamadan and similar cities. This methodology not only ensures a rigorous examination of pedestrian infrastructure but also contributes to broader academic discussions on sustainable urban planning and the socio-economic benefits of walkable streets. Hamadan's historical and cultural landscape serves as an exemplary case study, offering lessons for integrating modern urban planning with heritage preservation in non-Western contexts.

10.2.5 Scientific Background

Studies have explored the relationship between the built environment and walking behaviour, particularly in promoting sightseeing through walking streets. Luo et al. (2022) investigated urban residents' preferred walking street settings and environmental factors, highlighting the importance of the built environment, walking behaviour for public health and the effect of street environmental quality on walking behaviour (Luo et al., 2022). They identified three main street settings (community streets, waterfront paths and urban greenways) and examined the physical and aesthetic environmental factors that influence the selection of preferred walking streets based on the street environment. In the other research, a conceptual model representing the elements of internal and exterior environments is essential for assessing a city's tourism competitiveness (Cibinskiene & Snieskiene, 2015; Vlot-van Anrooij et al., 2020). The study found that internal and exterior surroundings have about identical weight coefficients (0.45 and 0.55), indicating their relative importance to the city's tourism competitiveness. Wayfinding information systems can be a cutting-edge strategy in the fight against over-tourism, suggesting that the intended study goal was accomplished (Mor & Dalyot, 2020). Scholars can apply a technique to evaluate the implications of walkability for tourists in the pre during, and post-development stages of community-based tourism by using themes and potential indicators (Jamhawi et al., 2023). The transdisciplinary Tourist Walkability Sustainable Tourism Impact on Residents (or TWS) research framework can be used to quantify the degree of walking potential of tourists in chosen tourist sites (Yeap & Liow, 2023). The relationship between walkability of neighbourhoods and pedestrian activity in the larger Tehran metropolitan region is examined. It shows net residential density, land use mix and street connectivity are the three characteristics that are considered, together with the socio-economic status of the populace at the statistical block level (Sakib et al., 2024). Reza Zadeh et al (2011) conducted a mental assessment of pedestrianization capability and its influential factors in urban neighbourhoods, exploring the relationship between walkability and the pleasantness of walking acknowledges that policies governing pedestrian path design play a crucial role in shaping the social fabric and constructive spirit of the city (Hosseini et al., 2016). The sense of belonging to a place and understanding one's living environment are pivotal in fostering the tranquillity and mental balance of city residents. In the other study investigates pedestrian behaviour in relation to residential and commercial areas, revealing that 56% of pedestrian potential is attributed to physical environmental factors and 53% includes cultural and social indices, with security being the most significant aspect among cultural-social factors (Boromandi, 2015).

10.3 Tourism and City

Tourism in the city refers to tourism activities that take place in an urban environment, attracting visitors for leisure and business purposes. Cities offer a concentration of cultural, architectural, technological, leisure and social attractions, making them appealing to tourists (Bouchon & Rauscher, 2019). However, there are different types of cities with varying population, size, functions and geographical locations, leading to differences in the economic contribution of tourism and its visibility within the urban landscape (Yang et al., 2022). The relationship between the city and tourism is complex and multifaceted. Urban tourism is not limited to sole cities but also includes other urban places that have been overlooked in research (Kaya and Keles, 2022). Additionally, tourism is not only a product of cities but also contributes to the urbanization of other areas such as seaside resorts and mountains (Coëffé & Stock, 2021). The development of city tourism is influenced by both endogenous factors related to the city itself and exogenous factors related to its socio-economic relations with the environment, including tourist traffic (Panasiuk, 2020). Pedestrian behaviour and walkability are important factors in the context of tourism. Studies have shown that pedestrian behaviour can be influenced by factors such as familiarity with the environment, road traffic rules and lighting conditions (Asriana, 2021). Designing pedestrian facilities and improving walkability in tourist destinations can contribute to creating attractive, accessible, safe and healthy environments for both tourists and residents (Gregoriades & Dimitriou, 2020; Hall & Ram, 2018). Walkability has been found to have positive impacts on the tourist experience, as it allows for exploration of urban spaces and contributes to sustainable mobility. However, the relationship between walkability and indicators of successful tourism, such as visitor numbers and reviews, has shown only weak correlations in some studies (Farkic et al, 2015). Pedestrians connect us to tourism by playing a crucial role in the tourist experience and contributing to the development of walkable cities. Pedestrian behaviour can be understood and modelled through agent-based simulation, which helps designers depict movement flow and design walkable spaces (Asriana, 2021).

Hassanshahi et al., (2023) showed in their research the importance of developing pedestrian tourism routes to enhance the walking experience in urban tourism destinations, with a focus on the historical region of Shiraz. This study utilizes a multi-criteria decision-making approach using Geographic Information System (GIS) for ranking street segments and applies techniques such as the Analytic Hierarchy Process (AHP) and the Order Preference Similarity to Ideal Solution (TOPSIS). The findings provide a systematic framework for planners and policymakers to make informed pre-investment decisions that promote walking and improve the overall tourism experience in urban areas.

The city of Shimla in India provides a prime example of a pedestrian street, known as the 'Mall Road', which offers a unique experience to tourists and can be studied as a model for pedestrianization in other destinations (Khan, 2014). Landmarks play a significant role in pedestrian orientation and can enhance the vacation experience for tourists, leading to more and longer walks (Millonig and Schechtner, 2025). Walkability is not only important for locals but also for tourists, as walking is an integral part of the tourist experience and contributes to sustainable and prosperous communities (Farkic et al, 2015). Ensuring the safety of pedestrians in crowded tourist areas is also a crucial task (ITF, 2012). It emphasizes the importance of tourism policymakers and entrepreneurs considering the effects of government policies, national culture and the individual personalities of both tourists and residents when developing strategies for sustainable tourism. These factors are crucial, as they can significantly affect the sacrifices and inequities faced within communities (Yeap & Liow, 2023).

Eigner and Bawole (2023) in their research believe that in Yogyakarta, Indonesia, the quality of pedestrian paths is crucial for connecting tourists to key attractions, although challenges such as narrow lanes and inadequate facilities remain. Walkability plays a significant role in fostering sustainable tourism by encouraging healthier lifestyles and fostering social sustainability. It facilitates connections among individuals within urban environments, thereby enriching the overall experience for tourists (Amen et al., 2023).

As Hassanshahi et al (2023) have pointed out walking improves the tourist experience by providing safe and convenient pedestrian routes, as observed in Shiraz, Iran, where a GIS-based approach was used to prioritize tourist routes based on criteria such as traffic safety and accessibility. In fact, the Walking Sustainable Tourism Impact on Residents (TWSTIR) framework highlights the triple benefits of walking, including economic, social and environmental impacts on the quality of life of the community (Yeap & Liow, 2023).

Sustainable tourism, as a new development paradigm, includes three essential aspects: environmental sustainability, socio-cultural sustainability and economic sustainability. These elements are interconnected and influence each other. Achieving sustainable tourism development requires refining development methodologies and establishing robust implementation systems to maintain a balance between tourism growth and environmental preservation (Choi, 2024).

Bello and Lambulira (2024) define sustainable tourism as the development of tourism in protected areas that focus on environmental conservation and the protection of natural resources while also addressing human welfare and the sustainable social and economic development of local communities. This type of tourism helps improve the economic status and cooperation of local communities by providing job opportunities, social development projects and income-generating activities (Choi, 2024) and (Bello & Lambulira, 2024).

Danarta et al. (2024) in their article show that sustainable tourism is defined by principles of nature and cultural conservation, which can support the development of halal tourism. These sustainability principles in halal tourism development create economic opportunities, enhance the quality of tourism services and strengthen Indonesia's position as a leading halal tourism destination.

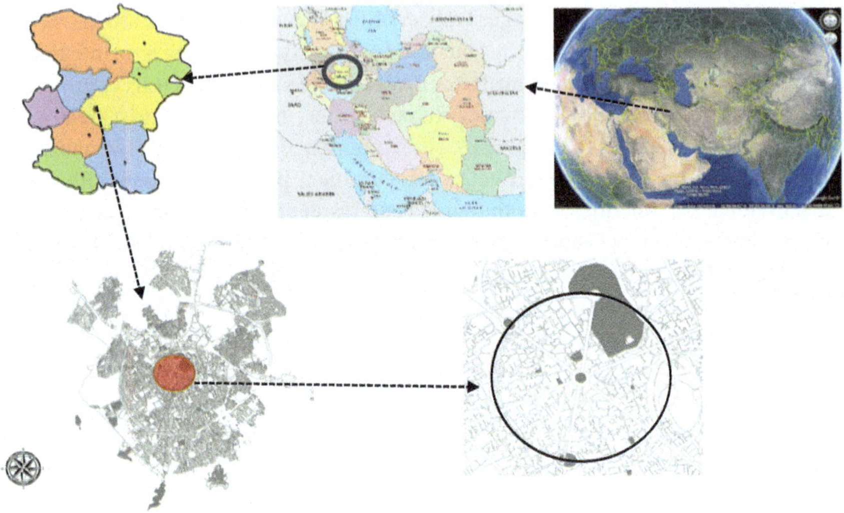

Fig. 10.1 Location of Hamedan in the Iran map and the study area

10.4 Understanding the Subject Matter and the Case Study

10.4.1 Historical Significance of Hamadan

Hamadan, located in western Iran, is one of the oldest cities in the country, with a history spanning over 2000 years. Known historically as Hegmataneh, meaning a gathering centre, it later became Ecbatana, the capital of the Median Empire, as noted by the Greek historian Herodotus. Over the centuries, Hamadan served as a capital for various dynasties, including the Achaemenid, Parthian and Sassanian empires, leaving behind a legacy of historical and architectural richness. Notable landmarks such as the Tomb of Avicenna, the Ganjnameh inscriptions and the Alavian Dome reflect this deep historical heritage. In 2006, the Islamic Consultative Assembly declared Hamadan the 'Capital of History and Civilization of Iran', further cementing its cultural significance. The city's historical relevance makes it a prime case study for exploring the interplay between urban infrastructure and tourism (Fig. 10.1).

10.4.2 Geographical and Urban Context

Situated on the slopes of Mount Alvand at an elevation of 1741 m above sea level, Hamadan is known for its mountainous terrain and cold climate, making it one of Iran's coldest cities. The city covers an area of approximately 62 square kilometres and has a population of around 560,000, ranking it as Iran's 13th most populous city.

Hamadan's urban layout reflects both historical and modern influences. Designed by the German engineer Carl Fritsch between 1921 and 1925, the city features a radial plan centred around Bu-Ali Square, where six major streets converge. This design facilitates connectivity but also creates challenges, such as traffic congestion due to population density. The juxtaposition of Hamadan's historical core with modern interventions makes it an ideal context for studying the dynamics of walkability and tourism (Fig. 10.3).

10.4.3 Tourism Potential and Challenges

Hamadan attracts approximately 300,000 tourists annually, drawn by its cultural and historical landmarks. However, compared to other prominent tourist destinations in Iran, the city's tourism potential remains underutilized. Factors such as limited infrastructure, inadequate promotion and heavy traffic in the city centre hinder its ability to fully capitalize on its cultural assets. Despite these challenges, Hamadan's proximity to significant landmarks and its pedestrian-friendly streets presents opportunities for sustainable tourism development. The city's historical core, characterized by a network of walkable streets and cultural nodes, offers a unique setting to explore how pedestrian infrastructure can enhance the visitor experience (Fig. 10.2).

10.4.4 Relevance to Walkability and Sustainable Tourism

Hamadan's blend of historical significance, cultural heritage and urban structure provides a compelling framework for investigating the socio-economic benefits of walkable streets. The city's radial design and dense network of pedestrian routes create an environment conducive to studying how accessibility impacts tourist behaviour and local engagement (Fig. 10.3). By analysing the interaction between pedestrian pathways and key landmarks, this research contributes to the broader discourse on sustainable tourism and urban planning in historical cities. Additionally, Hamadan serves as a valuable case study for understanding walkability in non-Western contexts, offering insights into how historical cities can balance modernization with cultural preservation. The findings have the potential to inform strategies for promoting sustainable tourism in similar cities worldwide (Figs. 10.4 and 10.5).

10.5 Spatial Analysis of Hamadan City Core

Hamadan, the 10th largest city in Iran, exhibits a distinctly monocentric urban structure. This pattern is evident from the results of the graph analysis using Angular Segment Analysis (ASA) applied to the OpenStreetMap (OSM) centreline data.

Figure 10.2 Tourist destinations

Although Hamadan's city core does not stem from a medieval circular layout, early twentieth-century modernization profoundly transformed the urban fabric, shaping its current configuration. The Segment Integration and Segment Normalized Angular Choice values derived from the ASA reveal the city's spatial dynamics in 2023 (Fig. 10.6).

The analysis highlights how the inner city, traditionally the historical and cultural origin of Hamadan, has evolved in response to modern urban growth. While the inner city no longer holds the highest integration and choice values, it remains a key area for urban interaction. This reflects the historical centre's adaptive response to modernization, maintaining a degree of spatial closeness in the context of the global city network. Further analysis of pedestrian segments within various radii (400, 800, 1200, 1600, 2000, 2400, 2800, 3200, 3600 and 4000 m) demonstrated that the most effective pedestrian flow occurs within the 2000 to 3600 m range, corresponding to a one-hour walking distance.

This finding underscores the importance of these spaces for tourism, as they align with the typical range of natural pedestrian movement. These results were validated by field observations and tracing studies, which identified significant pedestrian activity in the central areas of the city. Field surveys conducted in two key pedestrian spaces in Hamadan's inner city revealed a clear distinction between dynamic

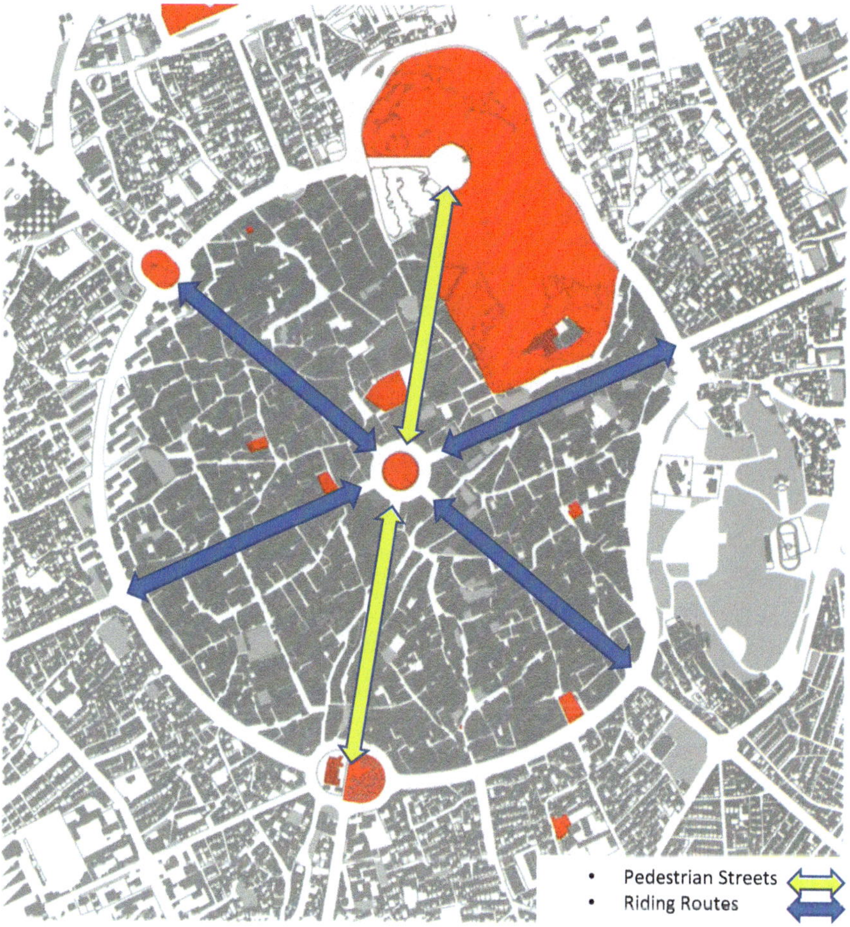

Figure 10.3 Main axes

and static spaces along pedestrian routes (Fig. 10.7). The city centre consistently emerged as the most integrated and historically significant area, confirming its role as a central hub for both locals and tourists. The tracing data further illustrated how the historical core continues to anchor pedestrian movement, despite the imposition of new urban layouts during the modernization of the twentieth century. The analysis of the city's pedestrian behaviour reveals how the twentieth-century transformations reshaped the inner-city fabric while preserving its central role in urban life. Figure 10.2 highlights how these changes influenced pedestrian patterns, showing that while the city centre no longer operates strictly as a neighbourhood or city-level hub, it maintains its interaction with major tourist attractions and urban functions. This reinforces the idea that the historical roots of Hamadan's fabric remain active, facilitating natural pedestrian movement and aligning with the principles outlined

Figure 10.4 Station points

by Space Syntax theory. The study also confirms that the pedestrian routes identified through Space Syntax analysis align closely with the natural movement of people in the city. The overlap between theoretical predictions and observed behaviours validates the hypothesis that the most integrated spaces in Hamadan are also the most frequently used by pedestrians. This supports the theory that integration and choice values provide reliable insights into pedestrian flow and interaction patterns in urban spaces. The findings of this study emphasize the resilience of Hamadan's historical core in adapting to modern urban growth while maintaining its significance for tourism and pedestrian activity. The results also highlight the importance of focusing on mid-range radii (2000–3600 m) in planning walkable spaces for urban tourism. These spaces, confirmed by both spatial analysis and field surveys, represent the critical zones where pedestrian infrastructure can have the greatest impact on accessibility, tourism and socio-economic vitality.

Figure 10.5 Busy roads

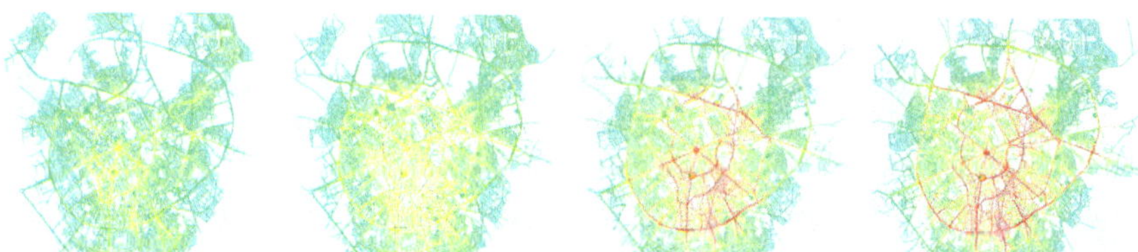

a) Int *r1200* b) Int *r2000* c) Int *r2800* d) Int *r3600*

Figure 10.6 Metric analysis for Integration in Hamedan city core

a) NACH *r1200* b) NACH *r2000* c) NACH *r2800* d) NACH *r3600*

Figure 10.7 Metric analysis for Normalize Angular Choice (NACH) in Hamedan city core

10.6 Conclusion

This study underscores the critical role of pedestrian streets in enhancing accessibility to tourist attractions within Hamadan's centralized urban fabric. The spatial configuration of the city, with its key pedestrian and vehicular axes, facilitates both local and tourist movement, reinforcing the importance of walkable spaces in urban tourism development. Analysis of choice maps within radii of 1600 to 3600 m revealed these zones as the most effective for pedestrian activity, aligning with natural movement patterns and increasing the likelihood of visitation to nearby tourist attractions.

The north–south pedestrian axis and the two main vehicular axes leading to the central square form a highly accessible and utilitarian structure, particularly within the 1600 to 2400 m range. Beyond these radii, the influence of the city centre diminishes, emphasizing the spatial centrality of Hamadan's core. These findings validate the hypothesis that pedestrian and tourist movements are interdependent and mutually reinforcing. The integration of tourist attractions with pedestrian streets enhances both accessibility and the distribution of visitors, contributing to a more vibrant and pedestrian-friendly urban environment.

Hamadan's centralized structure and surrounding belt of high pedestrian and vehicular traffic demonstrate the importance of integrated planning for supporting both local and tourist movements. The study highlights the potential of mid-range radii (2000–3600 m) as critical zones for urban tourism, where pedestrian infrastructure can have the greatest impact. These findings suggest that well-planned pedestrian streets not only improve access to tourist sites but also enhance the socio-economic vitality of urban cores.

The synergy between natural pedestrian flows and tourist attractions highlights the need for urban planners to prioritize walkability in historical cities. Effective pedestrian infrastructure can contribute to sustainable tourism by reducing reliance on motorized transport, preserving cultural heritage and fostering social interactions within the urban fabric.

One limitation of this study is the lack of precise data on the number of tourists visiting specific areas and their movement patterns relative to pedestrian pathways. Future research should incorporate real-time tracking technologies and visual analytics to capture accurate pedestrian flows and evaluate the effectiveness of pedestrian infrastructure. Additionally, the relationship between the spatial positioning of accommodations and tourist movement warrants further investigation. Network analysis that integrates accommodations with tourist routes could provide valuable insights into how these factors influence visitor behaviour and distribution.

To address these gaps, future studies could employ continuous tracking systems and GIS-based analyses to monitor pedestrian flows in real time. Exploring the integration of accommodation networks with pedestrian infrastructure could offer actionable recommendations for improving accessibility and enhancing the visitor experience. Finally, the findings of this study could inform urban tourism strategies in similar historical cities, emphasizing the global relevance of walkable streets in fostering sustainable tourism and urban development.

Acknowledgements This work is financed by Portuguese funds through FCT Fundação para a Ciência e a Tecnologia, I.P., under the Strategic Project with the references UIDB/04008/2020 and UIDP/04008/2020.

References

Amen, M. A., Afara, A., & Ahmad Nia, H. (2023). Exploring the link between street layout centrality and walkability for sustainable tourism in historical urban areas. *Urban Science, 7*(2), 67. Available at: https://doi.org/10.3390/urbansci7020067

Asriana, N. (2021). Pedestrian behavior for developing strategy in tourism area; agent-based simulation', *DIMENSI (Journal of Architecture and Built Environment), 48*(1), 65–74. Available at: https://doi.org/10.9744/dimensi.48.1.65-74

Bello, F., Lambulira, M. (2024). Pursuing Sustainable Tourism Development Through Wildlife Tourism: Local Residents' Perceptions. In L. S. Stone, & M. T. Stone (Eds.), *Wildlife tourism dynamics in Southern Africa* (pp. 115–130). Available at: https://doi.org/10.1007/978-3-031-57252-4_9

Boromandi, F. (2015). Rehabilitation of Lalezarzar Street as one of the historical-cultural center of Tehran. *Memarnet.*

Bouchon, F., & Rauscher, M. (2019). Cities and tourism, a love and hate story; towards a conceptual framework for urban overtourism management. *International Journal of Tourism Cities, 5*(4), 598–619. Available at: https://doi.org/10.1108/IJTC-06-2019-0080

Choi, Y. (2024). Active response for sustainable tourism. *Global Journal of Tourism, Leisure and Hospitality Management, 1*(3), 555–565. Available at: https://doi.org/10.19080/GJTLH.2024.01.555565

Cibinskiene, A., & Snieskiene, G. (2015). Evaluation of city tourism competitiveness. *Procedia—Social and Behavioral Sciences, 213*, 105–110. Available at: https://doi.org/10.1016/j.sbspro.2015.11.411

Coëffé, V., & Stock, M. (2021). Tourism as urban phenomenon and the crux of "urban tourism" (pp. 185–202). Available at: https://doi.org/10.1007/978-3-030-52136-3_11

Danarta, A., Pradana, M. Y. A., Abror, I., & Yahya, N. E. P. S. (2024). Encouraging religion and sustainable tourism as a conception of Indonesian halal tourism development. *RGSA—Revista de Gestão Social e Ambiental, 18*(7). Available at: https://doi.org/10.24857/rgsa.v18n7-073

Figner, & Bawole, P. (2023). Kajian Kualitas Walkability di Koridor Jalur Wisata Kota Yogyakarta: Studi kasus: Koridor Jalan Panembahan Senopati, Jalan Sultan Agung dan Jalan Kusumanegara. ATRIUM. *Jurnal Arsitektur, 8*(3), 211–222. Available at: https://doi.org/10.21460/atrium.v8i3.181

Farkic, J., Perić, D., Lesjak, M., & Petelin, M. (2015). Urban walking: Perspectives of locals and tourists. *Geographica Pannonica, 19*(4), 212–222. Available at: https://doi.org/10.5937/GeoPan1504212F

Gregoriades, A., & Dimitriou, L. (2020) Naturalistic analysis of tourist pedestrians' spatial cognition. *Advances in Tourism, Technology and Management*, 3–13. Available at: https://doi.org/10.1007/978-981-15-2024-2_1

Halcomb, E. J. (2019). Mixed methods research: The issues beyond combining methods. *Journal of Advanced Nursing, 75*(3), 499–501. https://doi.org/10.1111/jan.13877

Hall, C., & Ram, Y. (2018). Measuring the relationship between tourism and walkability: Walk Score and English tourist attractions. *Journal of Sustainable Tourism, 27*. Available at: https://doi.org/10.1080/09669582.2017.1404607

Hassanshahi, G., Soltani, A., Roosta, M., & Askari, S. (2023). Walking as soft mobility: A multi-criteria GIS-based approach for prioritizing tourist routes. *Frontiers of Architectural Research, 12*(2). Available at: https://doi.org/10.1016/j.foar.2023.09.001

Hillier, B., & Hanson, J. (1987). CambridgeCore_Citation_27Dec2023.
Hillier, B., Hanson, J., & Graham, H. (1987). Ideas are in things: An application of the space syntax method to discovering house genotypes. *Environment and Planning B: Planning and Design, 14*(4), 363–385. Available at: https://doi.org/10.1068/b140363
Hosseini, S. T., Ariyana, M., & Abroodi, S. M. (2016). Transport and urban traffic management in Tehran with economic view. *IUESA, 4*(15), 95–109.
ITF. (2012). *Pedestrian Safety, Urban Space and Health*. OECD Publishing. http://dx.doi.org/10.1787/9789282103654-en
Jamhawi, M. M., Zidan, R. J., & Sherzad, M. F. (2023). Tourist movement patterns and the effects of spatial configuration in a cultural heritage and urban destination: The case of Madaba, Jordan. *Sustainability (Switzerland), 15*(2). Available at: https://doi.org/10.3390/su15021710
Kaya, M., & Keleş, Y. (2022) Şehir Turizmi Kavramından Ne Anlıyoruz? Kuramsal Bir Çözümleme. *Afyon Kocatepe Üniversitesi Sosyal Bilimler Dergisi, 24*(2), 810–822. Available at: https://doi.org/10.32709/akusosbil.1018619
Khan, S. (2014). Walking the walk: An evaluation of pedestrian tourism on the 'Mall Road' Shimla. *International Journal of Hospitality and Tourism Systems, 7*(1), 1–12. Available at: https://www.researchgate.net/publication/327345191
Luo, Y., Ruggiano, N., Bolt, D., Witt, J.-P., Anderson, M., Gray, J., & Jiang, Z. (2022). Community asset mapping in public health: A review of applications and approaches. *Social Work in Public Health, 38*(3), 171–181. Available at: https://doi.org/10.1080/19371918.2022.2114568
Mehdizadeh, J. (2000). Concepts and principles of pedestrianization [in Persian]. *Shahrdari-ha*, 2(19), 0–0. https://sid.ir/paper/444677/fa
Millonig, A., & Schechtner, K. (2025). City tourism—pedestrian orientation behaviour. *Journal of Urban Planning and Development*. Available at: https://doi.org/10.3390/ijerph19094934
Mor, M., & Dalyot, S. (2020). Enriching walking routes with tourism attractions retrieved from crowdsourced user generated data. In *ISPRS Annals of the Photogrammetry, Remote Sensing and Spatial Information Sciences*. Copernicus GmbH (pp. 95–102). Available at: https://doi.org/10.5194/isprs-Annals-V-4-2020-95-2020
Panasiuk, A. (2020) Policy of sustainable development of urban tourism. *Polish Journal of Sport and Tourism, 27*(2), 33–37. Available at: https://doi.org/10.2478/pjst-2020-0012
Reza Zadeh, R., Zebardast, E., & Ltifi Oskoui, L. (2011). Perceptual measurement of neighborhood walk's ability and its influential factors in neighborhoods (Case study: Chizar neighborhood). *Urban Management, 28*, 297–313.
Sakib, N., Paul, T., Ahmed, M. T., & Momin, K. A. (2024). Investigating factors influencing pedestrian crosswalk usage behaviour in Dhaka city using supervised machine learning techniques. *Multimodal Transportation, 3*(1), 100108. Available at: https://doi.org/10.1016/j.multra.2023.100108
Van Nes Akkelies., & Yamu, C. (2021) Space syntax applied in urban practice. In *Introduction to Space Syntax in Urban Studies* (pp. 213–237). Cham. Available at: https://doi.org/10.1007/978-3-030-59140-3_7
Vlot-van Anrooij, K., Hilgenkamp, T. I. M., Leusink, G. L., van der Cruijsen, A., Jansen, H., Naaldenberg, J., & van der Velden, K. (2020) Improving environmental capacities for health promotion in support settings for people with intellectual disabilities: Inclusive design of the DIHASID tool. *International journal of environmental research and public health, 17*(3). Available at: https://doi.org/10.3390/ijerph17030794
Yang, Z., Zhang, S., Liu, J., & Sun, H. (2022). Network of tourism–industrial complex in cities: Typologies and implications through a critical literature review. *International Journal of Environmental Research and Public Health, 19*(9). Available at: https://doi.org/10.3390/ijerph19094934
Yeap, P.-F., & Liow, M. (2023). Tourist walkability and sustainable community-based tourism: Conceptual framework and strategic model. *International Journal of Tourism Cities, 10*, 78–104. https://doi.org/10.1108/IJTC-05-2022-0117

Chapter 11
Unveiling the Long-Term Value of Green Buildings for Self-Sufficient Units: A Life Cycle Cost and Budget Analysis in the Albanian Landscape

Julinda Keçi and Geron Rakipaj

Abstract The global emphasis on sustainability has urged an amplified interest in green buildings, which are designed to minimize environmental impacts and maximize economic and social advantages. Life Cycle Cost Analysis (LCCA) and Life Cycle Budget Development (LCBD) are fundamental in evaluating the economic viability and long-term sustainability of green buildings. These tools facilitate informed decision-making, cost optimization, and the creation of more sustainable and resilient structures, aligning with the overarching goal of sustainable development. Despite their importance, the application of LCC in the construction sector is constrained by the lack of standardized methodologies, tools, and benchmarks.

This study demonstrates the application of LCCA and LCBD to evaluate a green building in Tirana, Albania. The research methodology encompassed a thorough review of existing literature, an examination of green building case studies within the Western Balkans, and the creation of a region-specific framework. This framework facilitates the comprehensive identification and quantification of costs associated with the entire life cycle of green buildings, from design and construction to operation, maintenance, and end-of-life stages.

The findings reveal that green buildings are economically viable, with long-term savings from reduced energy and water consumption outweighing initial design and construction costs. Incorporating green building features from the design stage significantly influences life cycle costs and budgeting, particularly in reducing operation and maintenance expenses. Energy costs represent approximately 26.5% of the total life cycle budget, and reducing energy consumption is identified as the most influential factor in minimizing total life cycle costs.

The application of LCCA and LCBD in the Albanian context offers valuable insights for policymakers, building owners, and developers in the region, highlighting

J. Keçi (✉) · G. Rakipaj
Department of Civil Engineering, Epoka University, Tirana, Albania
e-mail: jkeci@epoka.edu.al

G. Rakipaj
e-mail: grakipaj16@epoka.edu.al

F. Naselli et al. (eds.), *Self-Sufficient Cities and Sustainable Habitats*, The Urban Book Series, https://doi.org/10.1007/978-3-032-10337-6_11

the economic and sustainability benefits of green buildings. The study contributes to the expanding literature on green building economics and lays the groundwork for further research and development of sustainable building practices in Albania, the Western Balkans, and beyond.

Keywords Green buildings · Life cycle cost analysis · Life cycle budget development · Sustainability

11.1 Introduction

The built environment plays a pivotal role in human society by providing shelter, enabling economic activities, and shaping urban landscapes. However, constructing, operating, and maintaining buildings significantly impact on the environment and incur substantial economic costs. In response to these challenges, the concept of green buildings has emerged as a transformative solution. These buildings incorporate self-sufficient features and sustainable practices to minimize environmental impacts, enhance energy efficiency, and promote the well-being of occupants. Self-sufficient green buildings go beyond traditional sustainability by integrating advanced systems that reduce resource dependency, optimize operational efficiency, and support long-term resilience.

A critical factor in the success of self-sufficient green buildings is detailed planning and management across their entire life cycle. Life Cycle Costing (LCC) provides a comprehensive framework to evaluate the total costs associated with a building, from initial design and construction to operation, maintenance, and decommissioning. Self-sufficiency is achieved through systems such as renewable energy generation, water recycling, and energy-efficient building envelopes, all of which require thoughtful integration from the earliest design stages. By extending the evaluation beyond financial metrics, Life Cycle Cost Analysis (LCCA) captures environmental and social dimensions, offering a holistic perspective on building sustainability.

Self-sufficient green buildings often entail higher upfront investments due to the integration of sustainable technologies and materials. However, these costs are offset by reduced operational expenses, such as lower energy and water consumption, minimal maintenance requirements, and enhanced occupant productivity. Features like double-skin facades, rooftop gardens, and rainwater harvesting systems illustrate the self-sufficient approach, enabling buildings to operate with reduced external resource inputs and enhanced resilience to environmental challenges. LCCA plays an important role in assessing these trade-offs, equipping stakeholders with the data needed to make informed decisions that align with long-term sustainability goals.

This study employs an integrated methodological approach to explore the self-sufficient aspects of green buildings in Tirana, Albania. Using LCCA and Life Cycle Budget Development (LCBD), the research examines a green building typology with a projected 40-year service life. The methodology encompasses a literature review, case study analysis from the Western Balkans, and the development of a tailored framework for assessing self-sufficient features. By quantifying costs across the building's life cycle, including energy, water, maintenance, and end-of-life phases, this study evaluates the economic viability and sustainability of self-sufficient green buildings.

The findings provide actionable insights for stakeholders in Albania and similar regions, demonstrating how self-sufficient green buildings can reduce resource dependency, lower life cycle costs, and contribute to sustainable urban development. This research underscores the importance of integrating life cycle planning into the adoption of self-sufficient green buildings, ensuring their role as a cornerstone for achieving sustainability and resilience in the built environment.

11.2 Literature Review

Green buildings have become a focal point in sustainable development, aiming to reduce environmental impacts while enhancing economic and social benefits. Life Cycle Cost Analysis and Life Cycle Budget Development are critical tools in evaluating the economic feasibility and long-term sustainability of green buildings. Several studies have highlighted the importance of LCCA in green buildings. For instance, Khasreen et al. (2009) emphasized that LCCA provides a comprehensive assessment of the total cost of building ownership, including initial construction, operation, maintenance, and disposal costs. Similarly, Ding (2008) argued that incorporating LCCA in the early design stages of green buildings leads to more informed decision-making and cost-effective solutions.

LCBD, on the other hand, focuses on the budgeting aspect, ensuring that the life cycle costs of a building are within the projected budget. Ramesh et al. (2013) demonstrated that LCBD helps in identifying cost-saving opportunities throughout the building's life cycle, thereby enhancing the overall sustainability of green buildings.

The benefits of green buildings are well-documented, including reduced energy consumption, lower environmental impact, and improved occupant health and productivity. A study by Kats (2003) found that green buildings offer significant energy savings, with LEED-certified buildings consuming an average of 25–30% less energy than non-certified buildings. Additionally, Miller et al. (2009) highlighted the positive impact of green buildings on occupant health, noting reduced absenteeism and increased productivity in green office buildings.

However, green buildings also present challenges, particularly in terms of higher initial costs and the complexity of integrating sustainable technologies. A report by the World Green Building Council (2013) noted that the perception of high upfront costs is a significant barrier to the adoption of green building practices. Moreover, the integration of renewable energy sources and advanced building materials requires specialized knowledge and expertise, which can be a challenge for construction teams.

While there is a wealth of studies on green buildings and their economic analysis through LCCA and LCBD, research specific to the Western Balkans and Albania remains limited. This gap is notable in the context of differing economic conditions, regulatory frameworks, and construction practices in these regions compared to Western Europe and North America, where much of the existing literature is focused. The unique climatic, cultural, and economic conditions of the Western Balkans and Albania demand tailored approaches to green building design, construction, and evaluation. For example, Vukmirovic et al. (2019) highlight the potential for green building development in the Balkans but note the scarcity of localized studies on LCCA and LCBD, which are essential for adapting green building practices to regional specifics. A study by Mustafa et al. (2018) on green building practices in Kosovo, a country in the Western Balkans, highlighted the lack of awareness and governmental support as major obstacles to the adoption of green building practices. This lack of region-specific research hinders the development of customized frameworks that could facilitate the broader adoption of green buildings in the area, suggesting a significant area for future investigation.

In conclusion, while the benefits of green buildings are well-documented, challenges related to costs, standardization, and regional adaptation remain. The literature review reveals a critical need for further research focusing on LCCA and LCBD in the context of the Western Balkans and Albania, where such studies could significantly impact the sustainable development of the built environment.

11.3 Green Building—Case Study

Green buildings are designed to be environmentally responsible and resource-efficient throughout their lifecycle, from planning and construction to operation and eventual deconstruction. According to the US Environmental Protection Agency (EPA), they prioritize practices that minimize environmental impact while maintaining functionality, durability, and comfort. Despite often having higher initial construction costs, the demand for green buildings continues to rise due to their ability to efficiently utilize renewable resources such as wind, water, and solar energy.

In the Western Balkans, the future of green buildings appears promising, driven by growing interest from both public and private sectors. Governments' commitments to energy efficiency and sustainable development, alongside international support and investment, are key factors fostering the adoption of green building practices in the region. In Albania, significant efforts are underway to incorporate green principles into urban development, particularly in Tirana, where a remarkable example of sustainable construction demonstrates the potential of self-sufficient design.

Located in the heart of Tirana, this eight-story reinforced concrete structure, completed between 2019 and 2021, spans 950 m^2 with a total floor area of 9100 m^2 (Fig. 11.1). It houses 56 residential flats and six commercial stores, all carefully designed with sustainability in mind. The building integrates various eco-friendly features aimed at reducing reliance on external resources while minimizing its environmental footprint. A rainwater harvesting system captures and stores rainwater for non-drinking purposes, such as irrigation and toilet flushing, which decreases

Fig. 11.1 Case study building located in Tirana, Albania

dependency on municipal water supplies and lowers water bills. Double skin facades, consisting of two layers of glass separated by an air cavity, enhance energy efficiency by improving natural ventilation, daylighting, and indoor air quality, significantly reducing the need for mechanical heating, ventilation, and air conditioning systems (Fig. 11.2).

The building also boasts a mini rooftop garden, which helps mitigate the urban heat island effect, manages stormwater, supports biodiversity, and improves the well-being of residents. Additionally, advanced water efficiency systems, including greywater recycling and pressure-reduction valves, significantly decrease water and energy consumption. Together, these self-sufficient features not only reduce the building's dependence on external utilities but also exemplify how sustainable design can balance environmental responsibility with practicality, comfort, and modern living standards.

11.4 Methodology

The research methodology involves a direct approach for Life Cycle Cost (LCC) analysis, employing the International Standard ISO 15686-5:2008 (2008) to identify LCC components and elements. The methodology includes estimating the building service life at 40 years, with a comprehensive analysis period from the building's construction in 2019 to its expected end-of-life in 2060. This period accounts for inflation and discount rates, alongside design, construction, operating, maintenance, and end-of-life costs. Data required for the analysis encompasses building service life, analysis period, inflation rate, discount rate, and various cost categories. This approach enables a detailed examination of the relationship between initial building

Fig. 11.2 Double Skin Facade wall applied in the case study building

features' costs and their long-term financial impact, providing a holistic view of a green building's lifecycle in the Albanian context.

11.5 Building Service Life

The assessment of a building's service life is a critical component in lifecycle cost analysis, involving the evaluation of the duration a structure is expected to achieve optimal performance. This consideration encompasses the durability of materials, architectural design, and systemic functionalities (ISO 15686-5:2008, 2008). A building with a longer service life tends to incur lower maintenance and replacement costs, thereby reducing overall lifecycle expenses. Adhering to the International Standard ISO 15686-5:2008, this study adopts a 40-year period as the recommended average service life for the green building under examination, aligning with the building's design specifications and international standards for sustainable construction practices.

11.6 Period of Analysis

The period of analysis defines the timeframe over which financial analysis of a structure's lifecycle costs and benefits is conducted. For green buildings, which prioritize long-term sustainability and efficiency, this period typically spans longer than for conventional buildings due to their higher upfront investments but lower operational and environmental costs over time (Armstrong & Walker, 2002). Research suggests adopting a 25–40-year analysis period for green buildings to effectively evaluate their net present value (NPV), as costs incurred beyond 40 years may have minimal impact (Kirk & Dell'isola, 1995). The international standard, ISO 15686-5:2008 (ISO 15686-5:2008,2008), advises against exceeding a 100-year period for similar reasons. Nonetheless, within sustainable development discourse, a comprehensive lifecycle cost analysis encompassing a building's entire lifespan is advocated (Swarr et al., 2011). Accordingly, this study will analyze the total lifecycle costs of the case study green building over a 40-year period, from its construction in 2019 to its projected end-of-life in 2060.

11.7 Inflation Rate

Inflation rate consideration is critical in the lifecycle cost analysis of green buildings due to its significant impact on the financial assessment's accuracy and viability. Inflation gradually increases the cost of goods and services, affecting construction, materials, labor, and operational expenses. To ensure the analysis reflects realistic future costs, it's essential to apply current inflation rates to predict cost increases over the building's lifespan, calculated based on accurate data and considering historical trends as well as future estimates (consumer price index). This study employs the Trading Economics inflation rate data for Albania from the past 26 years (Fig. 11.3) and the Consumer Price Index (CPI) (Fig. 11.4) and uses a 3.2% inflation rate for the case study's lifecycle cost calculations.

11.8 Discount Rate

The discount rate is crucial for determining the present value of future costs and benefits related to a building's lifecycle, reflecting the time value of money. This rate helps account for the reduced value of future earnings or expenses compared to present-day values, incorporating factors like inflation and opportunity costs. Public sector projects often use government-determined discount rates, as recommended by ISO 15686-5:2008 (ISO 15686-5:2008, 2008).

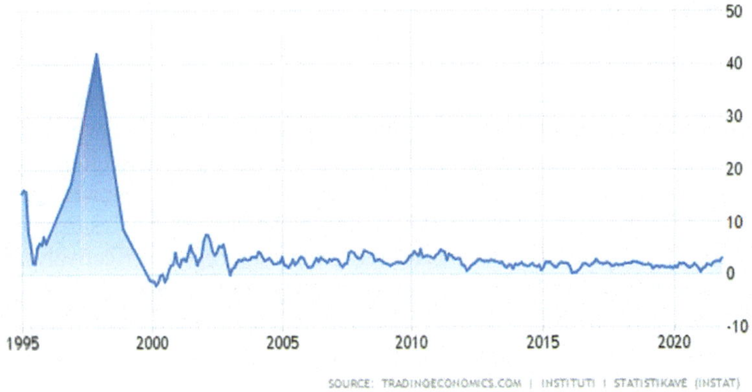

Fig. 11.3 Trading Economic Inflation rate of Albania from 1995 to 2021 (INSTAT, 2021)

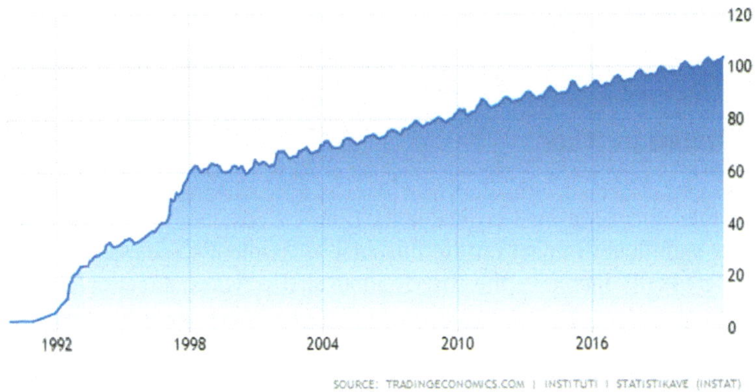

Fig. 11.4 Trading economic statistical data of Albania: CPI from 1989 to 2021 (INSTAT, 2021)

11.8.1 Design and Construction Cost

Design and construction costs form a significant part of a green building's lifecycle cost, covering all expenses from planning to construction completion. Accurate cost estimation is vital for evaluating a project's financial feasibility and sustainability. The case study building, with a gross area of 950m^2 constructed between 2019 and 2021, had its construction costs thoroughly calculated and presented in Euros for this analysis, with the total construction cost outlined in Table 11.1.

Table 11.1 Total estimated design and construction cost

Sr	Description of work		Amount in Euros
	Design cost		85,500
1	Amount of construction works carbines		900,809
2	Amount of construction works refined		1,326,724
3	Amount of parking construction works		68,989
4	Amount of electrical works		92,870
5	Amount of hydraulic + hydro-sanitary works		199,008
6	Fire protection		39,801
7	Amount of works 1 + 2 + 3 + 4 + 5 + 6		2,628,202
8	T.V.SH (government taxes)20%	20%	525,641
9	Total amount of works done including taxes		3,153,843

11.9 Building Operating Cost

Operating costs are important for lifecycle cost analysis of green buildings, encompassing ongoing expenses like energy, water, and sewage (ISO 15686-5:2008, 2008). Green buildings operational costs, particularly for energy and water, are typically lower due to their efficiency measures. This section details the methodology for estimating energy and water costs based on the building's consumption patterns and applies a 3.2% inflation rate to project these costs over the 40-year analysis period. Sewerage costs are also considered, highlighting the expense of wastewater treatment and its calculation based on water usage and sewer rates in Albania.

11.9.1 Energy Cost

Green buildings prioritize energy efficiency, resulting in lower energy consumption compared to conventional structures. The energy cost for the case study was calculated using a three-step methodology proposed by the European Green Building Council: estimating the building's projected energy consumption, calculating the cost of energy over the building's estimated lifespan, and discounting future energy expenses to their present value using the inflation rate as the discount rate. Based on a review of 30 billing records and considering the building's architecture, including materials, insulation, lighting, and HVAC systems, the projected energy consumption is determined to be 334,800 kWh/year, leading to an approximate annual energy cost of €26,282 (ERE, 2022). The discounted future energy expenses for the analysis period are presented in Table 11.2.

Table 11.2 Estimated life cycle energy cost for the case study

Estimated annual energy cost in (Euros/Year) for 40 Years					
Year	Annual cost	Year	Annual cost	Year	Annual cost
2021	26,282	2035	40,848	2049	63,488
2022	27,123	2036	42,155	2050	65,519
2023	27,991	2037	43,504	2051	67,616
2024	28,887	2038	44,897	2052	69,780
2025	29,811	2039	46,333	2053	72,013
2026	30,765	2040	47,816	2054	74,317
2027	31,749	2041	49,346	2055	76,695
2028	32,765	2042	50,925	2056	79,149
2029	33,814	2043	52,555	2057	81,682
2030	34,896	2044	54,236	2058	84,296
2031	36,013	2045	55,972	2059	86,994
2032	37,165	2046	57,763	2060	89,777
2033	38,354	2047	59,612	Total	**20,74,006**
2034	39,582	2048	61,519		

11.9.2 Water Cost

The total water cost was estimated using two data types: the building's total water usage for all purposes (m^3/year) and the water price in Albania (€/m^3). A review of 30 billing records, considering the water-saving measures incorporated into the building, concluded that the projected water consumption is 16,740 m^3/year, resulting in an approximate annual water cost of €10,513 (SHUK, 2021). The discounted future water expenses for the analysis period are presented in Table 11.3.

11.9.3 Sewerage Cost

Sewerage costs, associated with collecting and treating wastewater generated by the building, are a crucial aspect of the lifespan cost of green buildings. In Albania, sewerage services are provided by the Water Regulatory Authority (WRA), which also supplies, distributes, and controls freshwater. The total sewage cost was estimated using the estimated amount of wastewater generated (m^3/year) and the applicable sewerage rates (€/m^3). Based on the building's design and water usage, the average sewage quantity is estimated to be 15,210 m^3/year, leading to an approximate annual sewage cost of €2282 (WRA, 2021). The discounted future sewerage expenses for the analysis period are presented in Table 11.4.

Table 11.3 Estimated life cycle water cost for the case study

Estimated annual water cost in (€/Year) for 40 Years					
Year	Annual cost	Year	Annual cost	Year	Annual cost
2021	10,513	2035	16,340	2049	25,396
2022	10,849	2036	16,863	2050	26,208
2023	11,197	2037	17,402	2051	27,047
2024	11,555	2038	17,959	2052	27,912
2025	11,925	2039	18,534	2053	28,806
2026	12,306	2040	19,127	2054	29,727
2027	12,700	2041	19,739	2055	30,679
2028	13,106	2042	20,370	2056	31,660
2029	13,526	2043	21,022	2057	32,674
2030	13,959	2044	21,695	2058	33,719
2031	14,405	2045	22,389	2059	34,798
2032	14,866	2046	23,106	2060	35,912
2033	15,342	2047	23,845	Total	**8,29,618**
2034	15,833	2048	24,608		

Table 11.4 Estimated life cycle sewerage cost for the case study

Estimated annual sewerage cost in (€/Year) for 40 Years					
Year	Annual cost	Year	Annual cost	Year	Annual cost
2021	2282	2035	3547	2049	5512
2022	2355	2036	3660	2050	5689
2023	2430	2037	3777	2051	5871
2024	2508	2038	3898	2052	6059
2025	2588	2039	4023	2053	6253
2026	2671	2040	4152	2054	6453
2027	2757	2041	4285	2055	6659
2028	2845	2042	4422	2056	6872
2029	2936	2043	4563	2057	7092
2030	3030	2044	4709	2058	7319
2031	3127	2045	4860	2059	7553
2032	3227	2046	5015	2060	7795
2033	3330	2047	5176	**Total**	**1,80,081**
2034	3437	2048	5342		

The estimated operating costs for energy, water, and sewage highlight the importance of incorporating sustainable features in green buildings to minimize long-term expenses and contribute to overall sustainability.

11.10 Building Maintenance Cost

Maintenance cost is an essential component of a building's life cycle cost analysis. According to the International Standard ISO 15686-5:2008, maintenance costs for green buildings include labor, materials, and all other related costs associated with maintenance activities (ISO 15686-5:2008, 2008).

The estimated maintenance cost for the case study building in its first operational year (2021) is €9807 per year, as shown in Table 11.5. This estimate is based on the average maintenance and consumption charges from six operational green buildings in Albania. The total cost includes expenses for the maintenance and management team (€3722/year), as well as costs for maintenance services, security, cleaning, landscaping, healthcare, and sanitary products, all derived from the averages of these six buildings.

Using an average inflation rate of 3.2%, the maintenance cost of the case study building is projected for the next 40 years, from 2022 to 2060, as shown in Table 11.6. The total life cycle maintenance cost for the case study, accounting for inflation over the 40-year period, is estimated at €773,905. This amount represents the cumulative maintenance expenses the building will incur from 2021 to 2060.

Table 11.5 Estimated maintenance cost for one year (2021) year for the case study building

Estimated maintenance cost for 2021		
Sr #	Description of works	Costs (€/year)
1	Maintenance & Management Team Cost	3722
2	Maintenance service budget cost including lightning, firefighting, air conditioning, etc.	3210
3	Security services	1198
4	General cleaning of the building	780
5	Landscaping and gardening	657
6	Healthcare and sanitary products	240
Total estimated cost		**9807**

Table 11.6 Estimated maintenance cost for 40 years of the case study

Estimated annual maintenance cost in (€/Year) for 40 Years					
Year	Annual cost	Year	Annual cost	Year	Annual cost
2021	9807	2035	15,242	2049	23,690
2022	10,121	2036	15,730	2050	24,448
2023	10,445	2037	16,233	2051	25,231
2024	10,779	2038	16,753	2052	26,038
2025	11,124	2039	17,289	2053	26,871
2026	11,480	2040	17,842	2054	27,731
2027	11,847	2041	18,413	2055	28,618
2028	12,226	2042	19,002	2056	29,534
2029	12,618	2043	19,611	2057	30,479
2030	13,021	2044	20,238	2058	31,455
2031	13,438	2045	20,886	2059	32,461
2032	13,868	2046	21,554	2060	33,500
2033	14,312	2047	22,244	**Total**	**7,73,905**
2034	14,770	2048	22,956		

11.11 End-of-Life Costs

End-of-life costs encompass the expenses related to the dismantling, deconstruction, or demolition of a building at the conclusion of its service life. This critical aspect of lifecycle cost analysis accounts for inspection, demolition, waste disposal, and any associated disposal operations. Several demolition methods are employed for high-rise buildings, including mechanical, ballast (or implosion), deconstruction, and hybrid techniques (Pun et al., 2006). Mechanical demolition uses heavy machinery for building dismantling, while ballast demolition involves explosives to target key structural components for efficient collapse (Kibert & Languell, 2000). Deconstruction, a more labor-intensive approach, focuses on carefully dismantling buildings to recover recyclable and reusable materials, promoting sustainable development through reduced waste. The hybrid method combines mechanical and deconstruction techniques to optimize efficiency and sustainability.

The demolition process, particularly for high-rise structures, generates significant waste, necessitating careful management to uphold environmental sustainability. Research has established metrics for estimating demolition waste and labor requirements based on the building's area (Flanagan et al., 2005). For the case study in the subject, with a total floor area of 9100 m^2, the estimated demolition waste is 11,535 m^3, and the required labor amounts to 5460 hours. Using current Albanian rates for waste transportation and labor costs, the detailed end-of-life expenses are calculated as follows:

Site Clearing and Waste Transport: Based on a rate of 15.26 Euros/m^3, the total cost for waste transport is approximately 176,024 Euros.

Labor Costs: With a labor rate of 2.12 Euros/hour, total labor costs are estimated at 11,575 Euros.

Inspection and Supervision Costs: Accounting for 10% of the total demolition cost, this adds an additional 18,760 Euros.

Total End-of-Life Demolition Cost: The aggregate cost for the building's demolition is calculated to be 206,359 Euros.

If demolition were to occur in 2021, the total immediate cost would stand at 206,359 Euros. However, projecting this cost over 40 years with an annual inflation rate of 3.2%, the end-of-life cost is anticipated to rise to 704,906 Euros. This projection underscores the inflationary impact on future demolition costs and highlights the importance of incorporating these estimates into the lifecycle cost analysis for a comprehensive financial assessment up to the year 2060.

11.12 Results and Discussions

The Life Cycle Cost Analysis and Life Cycle Budget Development for the case study in the subject reveals a comprehensive overview of the building's economic performance over its 40-year lifespan. The analysis includes costs associated with design, construction, operation, maintenance, and end-of-life phases.

The total life cycle cost analysis of the eight-story structure with a floor area of 9100 m^2 reveals a comprehensive LCC of €7801,860 (Table 11.7). This cost encompasses all expenses associated with the building's life cycle, from construction to demolition, over 40 years.

The average LCC is calculated to be €857.3/m^2.

These results underscore the economic viability of green buildings, indicating that the initial higher investment in sustainable features can be offset by long-term cost savings.

The breakdown of the total LCC highlights that design and construction costs constitute 41.5% of the overall expenses, amounting to €3,239,343. Energy costs represent 26.5% of the total, at €2,074,005. Water and sewerage costs, essential for the building's 40-year service life, account for 13%, totaling €1,009,699. Building maintenance and end-of-life costs make up the remaining 10% and 9%, respectively, as detailed in Table 11.8.

The analysis highlights the significant impact of green building features on life cycle costs and budgeting. Sustainable design and construction practices contribute to reduced operational and maintenance expenses, enhancing the building's overall financial performance. Furthermore, energy cost and consumption emerge as critical factors in the life cycle budget. The building's energy-efficient design leads to substantial savings in energy costs, which significantly influence the building's life cycle cost and economic sustainability.

Table 11.7 Total estimated life cycle cost for the case study

Sr #	Year	Design and construction cost (€)	Building operating cost (€)	Maintenance cost (€)	Total annual LCC (€)	Cumulative LCC (€)
1	2019	1,619,672	0	0	1,619,672	1,619,672
2	2020	1,619,672	0	0	1,619,672	3,239,344
3	2021	0	39,077	9807	48,884	3,288,228
4	2022	0	40,327	10,121	50,448	3,338,676
5	2023	0	41,618	10,445	52,063	3,390,739
6	2024	0	42,950	10,779	53,729	3,444,468
7	2025	0	44,324	11,124	55,448	3,499,916
8	2026	0	45,742	11,480	57,222	3,557,138
9	2027	0	47,206	11,847	59,053	3,616,191
10	2028	0	48,717	12,226	60,943	3,677,134
11	2029	0	50,276	12,618	62,893	3,740,028
12	2030	0	51,885	13,021	64,906	3,804,933
13	2031	0	53,545	13,438	66,983	3,871,916
14	2032	0	55,258	13,868	69,126	3,941,043
15	2033	0	57,027	14,312	71,338	4,012,381
16	2034	0	58,851	14,770	73,621	4,086,002
17	2035	0	60,735	15,242	75,977	4,161,979
18	2036	0	62,678	15,730	78,408	4,240,388
19	2037	0	64,684	16,233	80,917	4,321,305
20	2038	0	66,754	16,753	83,507	4,404,812
21	2039	0	68,890	17,289	86,179	4,490,991
22	2040	0	71,094	17,842	88,937	4,579,927
23	2041	0	73,369	18,413	91,783	4,671,710
24	2042	0	75,717	19,002	94,720	4,766,430
25	2043	0	78,140	19,611	97,751	4,864,181
26	2044	0	80,641	20,238	100,879	4,965,059
27	2045	0	83,221	20,886	104,107	5,069,166
28	2046	0	85,884	21,554	107,438	5,176,604
29	2047	0	88,633	22,244	110,876	5,287,481
30	2048	0	91,469	22,956	114,424	5,401,905
31	2049	0	94,396	23,690	118,086	5,519,991
32	2050	0	97,416	24,448	121,865	5,641,856
33	2051	0	100,534	25,231	125,764	5,767,620
34	2052	0	103,751	26,038	129,789	5,897,409
35	2053	0	107,071	26,871	133,942	6,031,351

(continued)

Table 11.7 (continued)

Sr #	Year	Design and construction cost (€)	Building operating cost (€)	Maintenance cost (€)	Total annual LCC (€)	Cumulative LCC (€)
36	2054	0	110,497	27,731	138,228	6,169,579
37	2055	0	114,033	28,618	142,652	6,312,231
38	2056	0	117,682	29,534	147,216	6,459,447
39	2057	0	121,448	30,479	151,927	6,611,375
40	2058	0	125,334	31,455	156,789	6,768,164
41	2059	0	129,345	32,461	161,806	6,929,970
42	2060	0	133,484	33,500	166,984	7,096,954
43		End of life cost after service life, in 2060			704,906	78,018,60
Total LCC budget of Ish Profarma building					**7801,860**	

Table 11.8 Total life cycle budget summary for the case study

Sr. #	Description	Total LCC in €	Weight
1	Design and construction cost	3,239,343	41.5%
2	Building energy cost	2,074,005	26.5%
3	Water and sewerage cost	1,009,699	13%
4	Building maintenance cost	773,905	10%
5	End of life cost	704,906	9%
	Total LCC of building	**7801,860**	

11.13 Conclusions

This study provides critical insights into the economic and environmental sustainability of green buildings through Life Cycle Cost Analysis (LCCA), emphasizing the transformative potential of self-sufficient features. Design and construction costs accounted for 41.5% of the total life cycle cost, while energy costs represented 26.5%, underscoring the importance of reducing energy consumption to achieve significant long-term savings. The integration of advanced technologies such as double-skin facades, rainwater harvesting systems, and mini rooftop gardens demonstrates the economic viability of sustainable construction practices. These features, as applied in the case study, significantly lower resource dependency and operational expenses, with energy savings alone contributing substantially to financial feasibility.

Self-sufficient green buildings offer a promising future for sustainable infrastructure in Albania and the broader Western Balkan region. By addressing local challenges such as energy dependency and water scarcity, they align with national and regional sustainability objectives. The study underscores the importance of a comprehensive life cycle approach, from initial planning and design to end-of-life stages, to optimize resource use and cost efficiency. Features such as energy monitoring, water recycling, and modular construction enable buildings to maintain operational efficiency, resilience, and adaptability over time.

For policymakers, the findings advocate for incentives and regulations that promote green construction, ensuring broader adoption of self-sufficient building practices. Developers and building owners are encouraged to invest in sustainable technologies, which yield long-term cost savings and enhanced property value. Residents benefit from improved living conditions, reduced utility costs, and increased resilience to resource disruptions.

Future research should focus on refining LCCA methodologies, exploring innovative sustainable materials and technologies, and assessing the long-term performance of green buildings in diverse climatic and economic contexts. By addressing these areas, green buildings can advance as a cornerstone of sustainable urban development, contributing to the global goals of resource conservation and climate resilience. This study serves as a foundation for scaling self-sufficient practices in Albania, fostering a more sustainable and resilient construction industry.

References

Armstrong, J., & Walker, A., Green Building: n Project Planning and Cost Estimating, *Health, Comfort & Productivity,* 2002.

Ding, G. (2008). Sustainable construction—The role of environmental assessment tools. *Journal of Environmental Management, 86*(3), 451–464.

ERE, "www.ere.gov.al, " [Online]. Available: https://www.ere.gov.al/images/files/2022/02/18/Tarifat_dhe_cmimet_2022_18022022.pdf.

Flanagan, R., Jewell, C., & Norman, G. (2005). *Whole life appraisal for construction.* Blackwell Publishing Ltd.

INSTAT. (2021). "INSTAT " [Online]. Available: https://www.instat.gov.al/al/statistika/.

ISO 15686-5:2008. (2008). "Buildings, and constructed assets - Service-life planning - Part 5: Life-cycle costing, " International Organization for Standardization, Geneva, Switzerland

Kats, G.(2003). "The Costs and Financial Benefits of Green Buildings: A Report to California's Sustainable Building Task Force, " California Sustainable Building Task Force.

Khasreen, M. M., Banfill, P. F. G., & Menzies, G. F. (2009). Life-cycle assessment and the environmental impact of buildings: A review. *Sustainability, 1*(13), 674–701.

Kibert, C. J., & Languell, J. L. (2000). Implementing deconstruction in Florida: Materials reuse issues, disassembly techniques. *Economics and Policy.*

Kirk, S. J., & Dell'isola, A. J. (1995). *Life cycle costing for design professionals.* McGraw-Hill.

Miller, N., Pogue, D., Gough, Q., & Davis, S. (2009). Green buildings and productivity. *Journal of Sustainable Real Estate, 1*(1), 65–89.

Mustafa, M. A., Ismail, Z., & Keumala, N. (2018). Barriers to the adoption of green building practices in developing countries: The case of Kosovo. *Journal of Cleaner Production, 203,* 610–620.

Pun, S. K., Liu, C., & Langston, C. (2006). Case study of demolition costs of residential buildings. *Construction Management and Economics, 24,* 967–976.

Ramesh, T., Prakash, R., & Shukla, K. K. (2013). Life cycle energy analysis of buildings: An overview. *Energy and Buildings, 54,* 1–10.

SHUK. (2021). "SHUK, " [Online]. Available: https://www.shuk.al/tirane/tarifat.

Swarr, T. E., Hunkeler, D., Klöpffer, W., Pesonen, H.-L., Ciroth, A., & Brent, A. C., et al. (2011). Environmental life-cyclecosting: A code of practice. *The International Journal of Life Cycle Assessment, 16*, 389–391.

Vukmirovic, J., Maric, I., & Stojiljkovic, E. (2019). Green building development in the Balkans: Challenges and opportunities. *Journal of Cleaner Production, 212*, 317–329.

W. G. B. Council. (2013). The business case for green building: A review of the costs and benefits for developers. In *Investors, and occupants*. World Green Building Council.

WRA. (2021)."The water regulatory authority of Albania, ". [Online]. Available: https://www.wareg.org/articles/the-water-regulatory-authority-of-albania/.

Chapter 12
Lessons Learned from Sustainable Habitats and Self-sufficient Cities in the Euro-Mediterranean Context

Edmond Manahasa and Fabio Naselli

Abstract The limitations of natural resources on our planet necessitate serious consideration of efficient management strategies for urban and rural environments. Population growth and excessive consumerism have significantly contributed to the expansion of urban settlements, increasing energy demand, and reducing green spaces and natural habitats. Therefore, it is essential, as explored in this research epilogue, to identify lessons from best practices that can serve as advancements in developing sustainable habitats and self-sufficient cities. By referring to the global theoretical body, the study suggests the incorporation of innovative concepts such as resilience, circular economy and renewable energy provision into the curriculums of architecture and urban planning departments. Departing from a Euro-Mediterranean context, it also emphasizes people-centred development strategies, such as place-making, participatory design, co-planning and similar approaches, should be considered and implemented by local planning authorities to develop self-sufficient cities. Moreover, it highlights the relation between heritage buildings and their urban context and the urban elements of historical cities such streets or urban spaces, and their experiential aspects like vitality, walkability, socialization and qualitative visual experience, are essential features for creating sustainable habitats. In the current age of information technology, leveraging digital tools and technologies to understand public sentiment and engagement is vital for achieving sustainable and self-sufficient urban and rural habitats. In conclusion, while navigating ongoing challenges, embracing these approaches, tools and technologies can serve both as goals and valuable lessons for shaping the future of our habitats.

Keywords Sustainable habitat · Self-sufficient cities · People-centred approach · Physical · Non-physical features · Digital tools

---

E. Manahasa (✉) · F. Naselli
Department of Architecture, Epoka University, Tirana, Albania
e-mail: emanahasa@epoka.edu.al

F. Naselli
e-mail: fnaselli@epoka.edu.al

F. Naselli et al. (eds.), *Self-Sufficient Cities and Sustainable Habitats*, The Urban Book Series, https://doi.org/10.1007/978-3-032-10337-6_12

12.1 The Importance of Sustainable Habitats and Self-sufficient Cities in the Emerging Global Context

Factors such as raw resource exploitation, production and consumption cycles, population growth, climate changes, natural and human disasters, and the spread of conflicts have significantly impacted local, regional and global habitats (Correa, 2008, p. 17). Addressing these issues presents a high level of complexity in both research and real implementations, a complexity that requires being governed on a wider global scale. However, several 'promising' local -and localized- responses are emerging, focused on micro, small to medium-scale solutions, which look more reliable and manageable in achieving a needed sustainable balance between built and natural environments, between the anthropized and natural ones, and in promoting self-sufficiency as a practice to minimize the human ecological footprint (Syrovátka, 2020).

The integration of innovative concepts such as resilience, circular economy, and renewable energy becomes relevant in this scenario and in all the design fields. Additionally, people-centred development strategies that align with sustainability and self-sufficiency theories should be implemented by every local planning authority (Abubakar, 2022). The advanced information technology and the progress of ever-new digital tools can leverage these resources to manage the complexity of achieving sustainable and self-sufficient outcomes for governing the ongoing challenges at best (Das, 2024).

The relevance of these topics is underscored by the extensive literature available on both sustainable habitats and self-sufficiency subjects. Despite the number of existing studies, these subjects are still evolving, and conclusions are yet to be definitively established. There is a continuous need for understanding, testing, and discovering through real-life experiences and theoretical speculations. This exploration is useful for interpreting the effects of interrelations among diverse research focuses and, specifically, the integrations and relationships that emerge through genuine implementations. Both concepts of self-sufficiency and sustainable habitats involve multiple approaches and viewpoints, resulting in a complex picture that cannot be easily separated (Cappelli & Guallart, 2010; Guallart, 2014; Guallart 2022).

The present book is based on recent cases from six countries, focusing on sustainable, resilient, healthy and comfortable urban environments, considered vital for shaping future Sustainable Habitats and Self-Sufficient Cities. The research emphasizes the need for design and planning solutions that reduce environmental impact and are human centred. Recent implementations, although often early or experimental, underscore the importance of sustainability and self-sufficiency. These concepts, as highlighted by the UN-SDGs and EU-GD goals, involve reducing energy needs, minimizing soil consumption and waste production, promoting renewable energy self-production and encouraging recycling, reducing and reusing practices. In fact, these trends were discussed within design disciplines from the early twenty-first century, building upon insights from late twentieth century (Allaby & Bunyard, 1980).

The research presented produced a discourse on the subject departing from the emerging experiences of the presented works. It explored findings from case-by-case design or planning solutions, highlighting the global evolution of design thinking and how design approaches in architecture and urbanism have evolved in response to changes in predominantly urbanized habitats since 2016. Furthermore, through the presented works, the study promoted discussions and knowledge-sharing across design and planning disciplines by comparing various case studies, theoretical frameworks and initiatives from different regions around the Euro-Mediterranean Basin. The generated discourse from this research also aligns with the UN Agenda for the 2030 SDGs and the EU Green Deal (Koundouri et al., 2024). The effort is to track developments within trans-scalar design fields, exploring topics ranging from interconnected sustainable metropolises to green neighbourhood regeneration, from self-sufficient buildings to place-based self-sufficient micro-economies and from the roles/functions of parks in urban areas to historic urban habitat enhancement.

12.2 Achieving Sustainable Habitats and Self-sufficient Cities Through Planning and Design Tools in the Euro-Mediterranean Context

The works included in this book examined how design and planning can lead to greater self sufficiency in cities and more sustainable habitats for both the natural environment and human communities. The contribution of planning and design as tools to achieve sustainable habitats and self-sufficient cities is revealed by emphasizing the main outputs of each of the research as below:

Regarding self-sufficient city, a model as explained by Acierno and Pistone could be based on a medium sized city, considering the significance of 'size' as a model. They revealed that despite complexities, certain medium-sized cities like Santa Maria Capua Vetere (Italy) can display sustainable practices, highlighting their potential for resilience and self-sufficiency, and despite the obvious developmental obstacles they face.

On the other hand, conducting Life Cycle Cost and Budget Analysis Keçi and Rakipaj, reveal the long-term value of green buildings to achieve self-sufficient urban habitats. They emphasize the importance of application of LCCA and LCBD, for evaluation of a pilot green building and involve a research methodology developing a region-specific framework for the Balkans to identify and quantify costs associated with the entire life cycle of green buildings, from design and construction to operation, maintenance and end-of-life stages.

As for the sustainable habitats, especially the enhancement of a historic habitat in contemporary dynamics through urban projects and active participation to address 'correcting' top-down approaches is crucial. As proposed by Bellone and Trevisani, referring to Rome's Testaccio district, which is characterized by a compact historical fabric and consequently a strong local identity, when urban interventions are shared with residents, it helps to reduce the distorting effects of foreseen transformations.

Another way to consolidate the sustainable historical urban habitats is by revitalizing their place's memories, emphasizing 'literature, art and images' to enhance locals' urban identity and awareness as described by Eleftheraki. Especially, she emphasizes the potentiality of Landscape's 'mental mapping' for interpreting the local human cultural habitat in an Athen's neighbourhood and focusing on a basic characteristic of urban centres, their ability to evoke memories that fascinate us, as the city 'tells its story', through its buildings and structures, also highlighting its unprotected historical and cultural habitats.

In fact, the emergence of (un)sustainable abandoned and neglected habitats in rural and urban contexts contributes to the growth of a global 'third landscape', which relies on the exploration of further nature-based solutions, thus leading to sustainable habitats as proposed from Giametti and Merxhani in Albania. Especially departing from the issue of abandoned factories and drosscape in Albania, they defend the idea that the brownfields, disused factories and industrial landscapes may get a new meaning for a contemporary urban sustainable habitat, also emphasising the human health impacts of pollution.

The rising potential of 'Third Landscape(s)' for sustainable habitat is also revealed by Naselli and Giammetti who departing from the current visions for the next Tirana, featured by increasing high-rise buildings' new fabrics that disregard and overlap with the fragments of the historic fabrics, leading to the erasing of these original habitats such as inner public parts. As such, these original open spaces, transformed into a 'third landscape' born from this rapid growth of the formal city are evaluated as crucial to maintain an integrated, sustainable habitat.

Also to maintain sustainable habitats in historical urban context, the relationship between heritage buildings and their contexts amidst fast urban transformations is essential. Thus, highlighting heritage building's role as permanent fixtures within urban habitats to embed heritage legacy for sustainable neighbourhoods. As examined by Plyku Demaj et al. the assessment on the impact of urban changes on heritage buildings is crucial in understanding their up-to-date role in today's city and figuring out how such context can either preserve their historic values or overshadow them.

Similarly, the role of context and urban morphology in achieving sustainable urban habitats is very important as found by Pietrogrande, who refers to the town of Vigonza in Italy, due to its unique nature where the town hall and the church are in separate and distant places. Their solution recreates a sense of unity between these public urban domains within a more sustainable urban habitat based on the Italian school of typology through a balanced relationship between the new and the old in urban habitats, thus protecting and strengthening their cultural identity.

The role of public spaces to produce more sustainable multi-cultural urban habitat is revealed by Foadmarashi and Serdoura especially emphasizing their vitality, which relies on the ability to 'connect people'. They reveal that such vitality is present in the inner cities of Persian and Mediterranean areas due to their organic street networks that foster a sense of intimacy and community belonging. Departing from Urmia

Bazaar in Iran, they articulate its core role in serving as a complex city habitat, which beyond working as bustling centre of commerce and social exchange, is fostering connections, thus creating sustainable urban habitats.

Apart from public spaces in the historical inner cities' contexts in producing sustainable urban habitats, the role of walkable streets through enhancing the sightseeing in the attractiveness of tourist places is also essential as described by Asadpour et al. Furthermore, they found that the design and maintenance of footpaths play a crucial role in the attractiveness and accessibility of tourist places. This is because the well-designed footpaths influence the visual understanding of public space and develop behavioural aspects of suitable urban habitats for both local inhabitants and tourists.

As the subject of self-sufficient cities and sustainable habitats continues to evolve, we believe that it is essential that scholars need to stay informed about new issues and problem-solving approaches relevant to these topics. Thus, to certain extents, this book provided insights on effectively implementing design and planning tools for sustainable development encompassing city levels and its composing elements like buildings, neighbourhoods, streets or public spaces.

12.3 Lessons Learnt

In conclusion, this conference produced certain important lessons related to self-sufficient cities and sustainable habitats. Departing from this collection of 12 case studies, it exposed how design and planning applications can lead to greater self-sufficiency in our cities and more sustainable habitats for the environment and the human communities' future, which are summarized below:

- The medium-sized city is a possible model of a self-sufficient city, since this city scale can produce sustainable habitats and resilient practices for both the natural environment and local communities. Furthermore, economically self-sufficient neighbourhoods paired with digital technologies create self-sufficient habitats, much like green buildings that over time create a self-sufficient built environment.
- The maintenance of sustainable habitats in historical urban context that are subject to urban transformation operations remains a challenge: thus, inclusion of local people with participatory actions and also the evocation of place memories through mental maps and placemaking practices are crucial to preserve them.
- The intrinsic relation between heritage buildings and their urban context and also physical elements of the urban elements of historical cities such streets or urban spaces, and their experiential aspects like vitality, walkability, socialization and qualitative visual experience, are essential features that generate sustainable habitats.
- Finally, the study reveals that in a post-socialist context, both the neglected common public spaces (that have emerged due to the transformation dynamics of residential patterns) and the abandoned industrial landscapes contribute to

the evolution of new sustainable habitats, which rely on the concept of "Third Landscape".

References

Abubakar, A. (2022). Towards a human-centred approach for enhancing place prosperity: Defining and operating within the basic human needs. *Open House International, 47*(2), 190–206.
Allaby, M., & Bunyard, P. (1980). *The politics of self-sufficiency*. Oxford University Press.
Cappelli L., Guallart V. (2010). Self sufficient city. In *Envisioning the Habitat of the Future*. Aktar Publishers, N.Y., USA
Correa J. (2008). *Self-Sufficient Urbanism: A Vision of Contraction for the Non-distant Future*. Lulu publisher, USA
Das, D. K. (2024). Exploring the symbiotic relationship between digital transformation, infrastructure, service delivery, and governance for smart sustainable cities. *Smart Cities, 7*(2), 806–835.
Guallart V. (2014). The Self-Sufficient City. In *Internet has Changed our Lives but it hasn't Changed our Cities, Yet*. Aktar Publishers, N.Y., USA
Guallart V. (2022). *The Self-Sufficient Habitat*. Aktar Publishers, N.Y., USA
Koundouri, P., Alamanos, A., Plataniotis, A., Stavridis, C., Perifanos, K., & Devves, S. (2024). Assessing the sustainability of the European Green Deal and its interlin kages with the SDGs. *NPJ Climate Action, 3*(1), 23.
Syrovátka, M. (2020). On sustainability interpretations of the Ecological Footprint. *Ecological Economics, 169*, Article 106543.

Index

A
Accessibility, 13, 23, 29, 39, 107, 110, 143, 147–149, 152–157, 161–164, 167, 169, 172, 174, 199

B
Balanced urban development, 6
Bazaar, 6, 7, 143, 144, 147–158, 199
Brownfields, 6, 30, 34, 130, 198

C
Carbon footprint, 1
Cartography, 6, 57, 64, 80

D
Digital tools, 195, 196

E
Ecological consumption, 1
Ecological planning, 12

F
Fokionos Negri Street, 57, 59–64, 66, 67, 71, 72, 74, 75, 78, 80

G
Green buildings, 7, 177–180, 182–186, 188, 190, 192, 193, 197, 199
Green infrastructure, 1, 23–25, 28, 32–34, 53, 156, 157

H
Hamadan, 7, 161–164, 168–172, 174
Heritage buildings, 6, 113–117, 123, 195, 198, 199
Heritage impact assessment, 114
Historical habitat, 113, 116, 118, 123
Historic city, 46, 47, 98, 130, 132, 133
Historic urban landscape, 58, 59, 80

I
Identity, 5, 17, 23, 26, 33, 39, 42, 47, 57, 59, 67, 72, 84, 88, 93, 97–99, 107, 110, 115, 123, 125, 126, 129, 132, 151, 157, 162, 198
Industrial heritage, 83
Interwar urban heritage, 64
Iran, 5–7, 143–145, 147, 162, 167–169, 199

K
Kypseli, 57, 59, 61, 62, 64–66, 68, 71–79

L
Life cycle budget development, 7, 177–179, 190
Life cycle cost analysis, 7, 177–179, 188, 190, 192
Linear park, 43, 45, 46, 51, 52, 57, 59, 61, 63–65, 69, 77–80
Long-term viability, 1

M
Medium-sized city, 5, 16, 21, 197, 199

F. Naselli et al. (eds.), *Self-Sufficient Cities and Sustainable Habitats*, The Urban Book Series, https://doi.org/10.1007/978-3-032-10337-6

Memories, 26, 58, 59, 84, 132, 140, 198, 199
Minimized emissions, 1
Municipal planning, 18, 25, 33

P
Participatory process, 48
Pedestrian flow, 7, 143, 147, 148, 150, 154, 158, 161, 163, 164, 170, 172, 174
People-centred approach, 195, 196
Physical, 12, 13, 15, 21, 47, 58, 93, 97, 100, 101, 105, 107, 111, 115, 118, 132, 140, 146, 165, 199
Pollution, 4, 6, 13, 53, 91, 93, 198

Q
Quality of life, 1, 13, 41, 53, 97, 101, 111, 167

R
Revitalization, 106, 108, 109

S
Santa Maria Capua Vetere, 11, 16–25, 27–29, 33–35, 197
Self-sufficiency, 2, 3, 5, 12, 14, 16, 20, 30, 33, 53, 54, 178, 196, 197, 199
Self-sufficient cities, 1–5, 7, 11, 14, 195–197, 199
Spatial analysis, 103, 152, 156, 161, 163, 164, 169, 172, 174
Sustainability, 1–3, 6, 7, 13, 15, 23, 40, 41, 50, 53, 54, 58, 79, 113, 130, 132, 133, 137, 139, 156, 157, 161, 162, 164, 167, 177–180, 183, 184, 188–190, 192, 196
Sustainable city landscapes, 58
Sustainable environment, 1–4
Sustainable habitat, 1, 5–7, 39, 195–200
Sustainable planning, 11
Sustainable tourism, 162, 165, 167, 169, 174

T
Third landscape, 6, 84, 85, 91, 95–98, 102, 105, 110, 198, 200
Tirana next, 96, 110

U
Urban form, 14, 33, 100, 116, 130, 132, 140, 143, 147, 153
Urban morphology, 6, 30, 115, 125, 136, 198
Urban place, 132, 166
Urban project, 5, 39–41, 46, 47, 53, 54, 125, 132, 197
Urban regeneration, 11–13, 15, 17, 20, 22–25, 28, 29, 34, 53, 54, 84, 87, 132, 133
Urban resilience, 12, 22
Urban strategies, 41, 96
Urban unit, 113–121, 123, 124
Urmia, 6, 143–145, 147–158, 198

V
Vitality, 6, 41, 105, 143, 144, 146, 148, 149, 154–158, 162, 172, 174, 195, 198, 199

The manufacturer's authorised representative in the EU is Springer Nature Customer Service Centre GmbH, Europaplatz 3, 69115 Heidelberg, Germany. If you have any concerns regarding our products, please contact ProductSafety@springernature.com

Printed and bound by CPI Group (UK) Ltd, Croydon, CR0 4YY
17/09/2026
02218451-0007